THE NEW IMPERIALISM, VOLUME 2

Montreal, Quebec, Canada
2011

THE NEW IMPERIALISM

VOLUME II

INTERVENTIONISM, INFORMATION WARFARE, AND THE MILITARY-ACADEMIC COMPLEX

EDITED BY

MAXIMILIAN C. FORTE

ALERT PRESS

Montreal, Quebec, Canada
2011

Library and Archives Canada Cataloguing in Publication

Interventionism, information warfare, and the military-academic complex / edited by Maximilian C. Forte.

(The new imperialism (Montréal, Québec) v. 2)
Includes bibliographical references and index.
ISBN 978-0-9868021-1-9 (bound).--ISBN 978-0-9868021-0-2 (pbk.)

1. Imperialism. 2. Military research--Canada. 3. International relations. 4. Humanitarian intervention. 5. International relations and terrorism. 6. Propaganda, International. I. Forte, Maximilian C., 1967- II. Series: New imperialism (Montréal, Québec) ; v. 2.

JC359.I58 2011   325'.32   C2011-907014-6

Front cover image: An MQ-1C Sky Warrior unmanned aircraft system sits dormant in a hangar. This Sky Warrior aircraft was deployed in support of "Operation Iraqi Freedom". (Photo: Sgt. Travis Zielinski, 1st ACB, 1st Cav. Div., USD-C. In the public domain.)

Back cover image: Camp Taji, Iraq—Under the cover of night, an AH-64D Apache attack helicopter from the 1st Air Cavalry Brigade, 1st Cavalry Division, departs the flight line to conduct operations in support of "Operation Iraqi Freedom", Dec. 2, 2009. (U.S. Department of Defense photo. In the public domain.)

© 2011 Alert Press
1455 de Maisonneuve Blvd., W.,
Montreal, Quebec, Canada, H3G-1M8
www.alertpress.net

Printed in the United States of America

CONTENTS

FIGURES

PREFACE

The impact of WikiLeaks, and the politics of secrecy and national security embedded into information of geopolitical value, became prominent during late 2010 and the start of this seminar in early 2011, the second in our series. Then the so-called "Arab Spring" exploded, with some going as far as speaking of the Tunisian and Egyptian upheavals as revolutions, along with large protests and violent repression in Yemen, Syria, Bahrain, and Libya, and protests in Jordan and Algeria as well. U.S. foreign policy seemed both defensive and aggressive, hesitant and manipulative, with clear intentions to control events as much as possible in what might have been seen as a veritable new wave of decolonization against U.S. supported dictatorships. From there we then witnessed the implementation of one of the subjects of our seminar, the "Responsibility to Protect" (R2P), the phenomenon of allegedly "humanitarian" intervention, and the start of the NATO war against Libya, presumably to protect civilians with a "no-fly zone," yet quickly advancing towards its actual goal, regime change. In fact, discussion of Libya recurred throughout this seminar. In the meantime, the war in Afghanistan continued with as much Afghan resistance as ever, and Canada began to withdraw from combat operations in Kandahar—at the same time, we began to hear less of the glowing endorsements of "counterinsurgency" and "winning hearts and minds," and much more about air strikes and night raids. Haiti, as an international protectorate of the United Nations, occupied by thousands of foreign troops, saw the spread of a massive and lethal cholera epidemic brought by those troops, bringing their presence into sharp focus in connection with sexual assault on women and children, prostitution, and the UN Mission's continued display of arrogance toward Haitians. We also studied the U.S. Army's Human Terrain System, which saw changes during this past year with the resignation/firing of its top personnel, including anthropologist Montgomery McFate—and yet the program was to be expanded, with more overtures to universities, and more states supposedly interested in adopting the program for their own militaries. While we also examined "militainment"—the marriage between the military and the entertainment and news media—we witnessed the fact that some of the most popular movies to come out recently, such as *Avatar* and *District 9*, and strikingly different films such as *Buried*, went decidedly against the triumphalist, imperial gain of military worship. All of these events and issues largely formed the immediate context in which our seminar took place, with wide-ranging discussions touching on subjects as disparate as security theatre in airports to the role of NGOs in Haiti.

In many ways this second volume is different from the first, reflecting the different interests of participants, the context as described above, and a seminar structure that continues to evolve. The seminar is now much more geared toward publication, allowing greater length for each paper, but also increased expectations of more advanced work. In future seminars, even more time will be devoted specifically to the preparation of papers, and less on other tasks. At this point, the seminar is in a stage of transition. Both the content and orientation will likely change significantly before it is offered again after 2013, tentatively under a title with greater longevity: *Empire*.

As a companion to the chapters, we have included once more appendices that add to the depth of the papers, by offering primary materials, as well as classic publications now in the public domain. The function of the appendices is thus to provide a reader or compilation that complements and adds to the contributors' chapters, offering a further bonus to those acquiring this text.

INTERVENTION, INFORMATION, IDEOLOGIES, AND INDUSTRY: THE NEW IMPERIALISM AND ITS REFRACTIONS

Maximilian C. Forte

In the first volume, I provided a conceptual, theoretical, and historical overview of "the new imperialism," that staked out what have been some of the relatively novel and significant transformations of global geopolitics and political economy over the past 20 years, that is, since the end of the Cold War, that could justify us speaking of a new imperialism. The point was not to dismiss the considerable continuities, and repetitions of past periods and phenomena (which include what at the end of the nineteenth-century was also called "the new imperialism" in Britain), but rather to shine a light on the particularities of the present. Having done that, I will not be repeating that material here. Instead I wish to highlight the specific refractions of the new imperialism through four distinct angles as presented in the contributions to this volume: intervention, information, ideologies, and industry. While the chapters in this volume can be read in any order, and tend to stand alone as contributions (with occasional and diverse overlaps), for the purposes of this introduction I have organized them according to these broad thematic lines.

Intervention is of course a dominant concern in the study of imperialism, involving the direct projection of political and military power in order to significantly transform another society and to bend it to the will of the invader and occupier. In this volume, the questions raised by the seminar participants, around this topic, included: What are the prospects for humanitarian internationalism under imperial conditions of (im)possibility? When it comes to interventions conducted under the pretext of humanitarianism, does the U.S. target specific countries, and if so, what are the characteristics of those countries? What can we learn about "hard power" from the two wars against Iraq? How was NATO's war in Libya justified according to a broad humanitarian myth?

Information became a more prominent, critical battleground in the conflict between the national security state and civil society, between the imperial state and the media, especially with the rise to prominence of WikiLeaks with its release of hundreds of thousands of files leaked from within the U.S. military, intelligence, and diplomatic establishments. While WikiLeaks has been widely lauded, and has attracted a very large mass of ardent supporters—which in themselves are facts that speak to imperialism's crisis of legitimacy and the failures of U.S. "soft power"—critical questions remain to be addressed. Participants asked what impact WikiLeaks had in changing U.S. foreign policy. Has WikiLeaks lived up to its promise of "opening governments"? And then there were more specific, empirical questions raised elsewhere in the volume, such as: what do we learn about torture and U.S. war crimes from WikiLeaks' Iraq War Logs?

Ideological battles are inevitably enfolded in all of the performances of national security and the articulations of principles and policies of intervention. Yet, in this volume, the spotlight falls squarely on ideology itself, in the form of national rhetoric. We thus asked: What is the role of national rhetoric in the mutual antagonism between Iran and the U.S.? More broadly: Why does the issue of moral hypocrisy matter when it comes to interventions that some justify on humanitarian grounds? Do we expect such interventions to be morally pure, uncontaminated by other motives, or does hypocrisy really matter in shining a light on the workings of soft power, in putting public diplomacy to the test, in challenging the allegations and assertions found in various "strategic communications" and psychological operations? Others asked: How does one align torture with the defense of liberal democracy? All of these questions underscored the importance of ideology and moral precepts, not to be carelessly dismissed or diminished as if only practice mattered and not the theory that informs or justifies it.

Industry in this volume is a shorthand way of referring to the military-industrial-academic complex. This is a major theme in the seminar as a whole, focused on the recruitment of the social sciences, and the academy more generally, by the national security state. We asked: What are the processes, patterns, and agents behind the militarization of university research in Canada? What is the nature of the militarization of the academy—how has it in fact changed, and changed toward what? And we ask: Can academic research be critical and ethical when funded by the military and private defence contractors?

Now let us turn to some of these refractions as played out within the chapters and appendices included in this volume. The chapters in this volume are discussed in terms of the four main themes of the volume, and not necessarily in the order in which they appear in the table of contents.

Intervention

Corey Seaton's chapter, "Hard Power and Iraq: Destabilization, Invasion and Occupation," clearly focuses on some of the more direct and coercive aspects of U.S. foreign policy. Military intervention remains a persistent defining feature of imperialism, whether "old" or "new," and for this reason it will appear as an enduring feature of our discussions. (Propaganda, on the other hand, has been sufficiently developed, crafted, and advanced over time, so that it still merits—to some extent—inclusion as a defining feature of the "new" imperialism.) It is worth noting that, "when asked about the importance of America's soft power....then-Secretary of Defence Donald Rumsfeld claimed not to know what it was" (Nye quoted in Lock, 2010, p. 32). Seaton's chapter provides an alternate viewpoint on power, complementing Sabrina Guerrieri's chapter with its focus on soft power in the "war of words" between Iran and the U.S. (more below).

Seaton makes the basic and yet astute point that if the U.S. applied the doctrine of pre-emptive war evenly, to all those states that might pose a threat to American interests (however defined), then the U.S. would be invading countries all the time. But it does not. Why not? This is the first question Seaton addresses. He notes that "countries that are targets for pre-emptive war tend to have several characteristics: a) they must be virtually defenseless; b) they must be important enough to be worth the trouble; and, c) they must be able to be portrayed as the ultimate evil and threat to U.S. survival". It is in this frame that Seaton discusses "hard power" and U.S. global dominance, noting that hard power has been defined as resting on the capacity to get others to change their positions, either via threats of force (or actual force), or the use of "incentives".

What Seaton describes is a series of actions, from sanctions to outright invasion and occupation that have effectively destroyed Iraq over the past two decades, with millions killed from a combination of sanctions, the invasions, and the ensuing sectarian conflicts, and many more millions displaced in a nation that had little more than 25 million people to begin with. Far from achieving either stability, security, or democracy, U.S. and UN actions have produced the exact opposite. And all of this, based on bald lies and fundamentally illegal intervention.

Jessica Cobran in her chapter, "Humanist or Imperialist? Humanitarian Interventionism in the Post-Cold War Era" faults humanitarian interventionism by military means as going beyond just protecting civilians. Cobran devotes considerable attention to the question of "saving lives," one of the key selling points of what some call "humanitarian intervention," and others more appropriately call "military humanism". Perhaps we are wrong for taking the claim so seriously, inadvertently legitimating the "saving lives" narrative by taking it as a legitimate point of contention. In my own chapter, I describe just how far beyond protecting civilians such intervention can go in the case of Libya, where one cannot build a credible and coherent case that protecting civilians in any way accurately describes NATO's intervention.

Cobran examines some of the fallacies of humanitarianism under imperial conditions: the moral delusions, the hijacked hopes, the power plays that work to create one situation of human rights violations as catastrophic and in need of immediate attention, and the other as acceptable, strategically delicate, or not worthy of international interest. If humanitarianism on the international level is to have any chance for a future, and for credibility when states are involved, the outstanding issues of hypocrisy and selectivity will need to be addressed, and remedied, or the rhetoric of "protection" will have to be dropped altogether. As Cobran notes, selectivity is an unshakeable feature of the current practice of humanitarian interventionism, and she rightly asks why some cases seem to require a seemingly interminable amount of deliberation and hesitation, and other cases witness a rush to war. Perhaps here Corey Seaton provided some clues toward an answer, in focusing on the characteristics of the countries which the U.S. chooses to invade and occupy. To the extent that he is correct, humanitarian interventionism becomes a mere fig leaf for U.S. geopolitical domination, which is where Cobran's argument also tends to lead.

If there were critical lessons to be learned from the Iraq, Afghanistan and Kosovo wars, they seem to have undergone a process of state-sanctioned, mass-mediated un-learning. This is where my chapter at the end comes in, in furnishing a critical overview of "the top ten myths" of the war against Libya—a counterpoint to other articles that also refer to "the top ten myths" but for the purpose of supporting the war. In many ways this is a current version of some of the processes described by Seaton, with the addition of a heavy emphasis on "humanitarianism". Far from humanitarianism in Libya, what we have often seen NATO engaged in was, among other things, the direct targeting of civilians and civilian facilities in a manner that meets most official definitions of terrorism used by Western governments.

Terrorism. But who is the terrorist? "The U.S. military and the Bush administration among the other groups of individuals who approved and imposed the malicious and immoral acts on detainees are terrorists"—this is one of the arguments made by Natalie Jansezian in her chapter, "Torture and the Global War on Terror". If there was ever a deployment of "hard power" in the act of foreign military intervention, torture becomes its most emblematic form, so intimate, personal, and painful that it causes widespread shock and revulsion, not least when it is defended in the names of democracy and freedom. Jansezian's chapter discusses and analyzes the torture tactics used by U.S. personnel in the so-called "global war on terror," for which she provides some introductory background. From there she proceeds to briefly discuss the torture scandal of Abu Ghraib, Iraq, which was prominently featured in our first volume (Siddiqui, 2010). She also examines the forms of torture that have been used and how they have been used against detainees—important especially when administration officials in the U.S. government chose to quibble about this or that technique as not meeting their definition of "torture". Jansezian also raises the fact that although multiple reports as well as photographs indicated that the U.S. did, in fact, use torture in an attempt to extract information from detainees, "the Bush administration knowingly allowed and authorized the use of torture during the process of interrogation". She thus focuses on the Geneva Convention Relative to the Treatment of Prisoners of War and the Convention against Torture and Other Cruel, Inhuman, or Degrading Treatment or Punishment.

Torture, I note, continues to be selected as salient topic by seminar participants, some of whom, like Jansezian, see this as a key feature of the new imperialism. At least some of the intellectuals of the new imperialism, such as Michael Ignatieff (2004), warrant this consideration in their defense of the harsh tactics of the "carnivore" in defence of democracy, as "the lesser evil". What Ignatieff is doing of course is to implicitly affirm that states do evil, and should use torture, but without the intellectual honesty of Jansezian who describes this as *terrorism*.

More illuminating than Ignatieff in this regard is the work of Noam Chomsky on terrorism in international relations, and as defined and practiced by the U.S. (see Chomsky, 1991, 2002b, 2002c). Chomsky (2002b, ¶ 3) points out that "terrorism" is defined in U.S. Army manuals as, "the calculated use of violence or threat of violence to attain goals that are political, religious, or ideological in nature...through intimidation, coercion, or instilling fear" (US Army Operational Concept for

Terrorism Counteraction [TRADOC Pamphlet No. 525-37], 1984).
Chomsky (1991, ¶ 6) also quotes the official United States Code for its
definition of terrorism:

> "'act of terrorism' means an activity that—(A) involves a violent act or
> an act dangerous to human life that is a violation of the criminal laws
> of the United States or any State, or that would be a criminal
> violation if committed within the jurisdiction of the United States or
> of any State; and (B) appears to be intended (i) to intimidate or coerce
> a civilian population; (ii) to influence the policy of a government by
> intimidation or coercion; or (iii) to affect the conduct of a
> government by assassination or kidnapping".

In light of these definitions, Chomsky considers the U.S. bombing
of Afghanistan, blocking imports of food from Pakistan, and older cases
such as mining Nicaragua's harbours, for which the U.S. was found
guilty of state terrorism in 1986 by the International Court of Justice in
The Hague. He argues that, "it would be hard to find anyone who accepts the doctrine that massive bombing is the appropriate response to
terrorist crimes—whether those of Sept. 11, or even worse ones," a position that must follow "if we adopt the principle of universality: if an action is right (or wrong) for others, it is right (or wrong) for us" (2002c, ¶
5). Those who would disagree with that position "do not rise to the
minimal moral level of applying to themselves the standards they apply
to others" and "plainly cannot be taken seriously when they speak of
appropriateness of response; or of right and wrong, good and evil"
(2002c, ¶ 5). To do so would be to plainly take what Chomsky (1991, ¶
1) calls the *propagandistic approach* to the concept of terrorism, where
"we begin with the thesis that terrorism is the responsibility of some officially designated enemy. We then designate terrorist acts as 'terrorist'
just in the cases where they can be attributed (whether plausibly or not)
to the required source; otherwise they are to be ignored, suppressed, or
termed 'retaliation' or 'self-defence'". It would seem that this is the approach popular with bellicose public commentary in the U.S., which
protests that there is no "moral equivalence" and no justification for
"relativism" when it comes to weighing U.S. violence against the violence of those who fight back. Of course not: this is pure American ex-
ceptionalism, where the U.S. possesses some absolute, divine right to
behave however it wants, to be held above judgment, and to be praised
and admired without question. *When you do it, it's bad; when we do it, it's
good. When you fight, it's always offensive; when we fight, it's always defensive.*
It is a position we should know and understand, and not one that we
need to respect—although as propaganda designed to justify heinous acts
of massive brutality, we need to seriously examine it. Chomsky quotes

Michael Stohl who observes, "that by convention—and it must be emphasized *only* by convention—great power use and the threat of the use of force is normally described as coercive diplomacy and not as a form of terrorism," though it commonly involves "the threat and often the use of violence for what would be described as terroristic purposes were it not great powers who were pursuing the very same tactic" (1991, ¶ 2).

In the case of the U.S. bombing of Afghanistan, given the principles established, Chomsky argued that there is no way of seeing this as anything other than terrorism:

> "On October 12th, a couple of days after the bombing started, Bush publicly announced to the Afghan people that we will continue to bomb you, unless your leadership turns over to us the people whom we suspect of carrying out crimes, although we refuse to give you any evidence. That's probably because they don't have any. And we dismiss without comment the offers of your leadership for negotiations about extradition.

> "Notice that is a textbook illustration of international terrorism, by the U.S. official definition. That is the use of the threat of force or violence, in this case extreme violence, to obtain political ends through intimidation, fear and so on. That's the official definition, a textbook illustration of it. Three weeks later, by the end of October, the war aims had changed. They were first announced as far as I can find out, by the British Defense Minister, Sir Admiral Boyce. Admiral Boyce informed the Afghan population that we will continue to bomb you until you change your leadership. Well, that's an even more dramatic illustration of international terrorism, if not aggression. And that was the goal that was followed. This had nothing to do with finding the criminals and bringing them to justice". (2002a, ¶ 14-15)

Asked to distinguish between the terrorism of Osama Bin Laden and what Chomsky says is U.S. terrorism, this is how Chomsky responded:

> "It's very simple. If they do it, it's terrorism. If we do it, it's counter-terrorism. That's a historical universal. Go back to Nazi propaganda. The most extreme mass murderers ever. If you look at Nazi propaganda, that's exactly what they said. They said they're defending the populations and the legitimate governments of Europe like Vichy from the terrorist partisans who are directed from London. That's the basic propaganda line. And like all propaganda, no matter how vulgar, it has an element of truth. The partisans did carry out terror, they were directed from London. The Vichy government is about as legitimate as half the governments the U.S. has installed around the world and supports, so yes, there was a minor element of truth to it, and that's the way it works. If somebody else carries it out, it's terror. If we carry it out, it's counter-terror". (2002a, ¶ 19)

Interestingly, contemporary state propaganda about "counter-terrorism" is not the only discourse that echoes that of the Nazis. In Appendix F, "Adolf Hitler, to the assembly in the Reichstag on 01 October 1 1938, following the German invasion of Czechoslovakia," we also see the articulation of principles of humanitarian interventionism, of the kind invoked in the NATO war against the state of Libya. As military intervention continues to be a prominent concern in this seminar, appearing in the chapters by Seaton, Jansezian, Guerrieri, Cobran, and my own, and given our persistent return to discussing and debating the intervention in Libya, I have also included Appendices G and H, with two important and conflicting statements on the right of intervention in Libya.

WikiLeaks' revelations about the Iraq War, through its release of nearly 400,000 U.S. documents, also figure prominently in Jansezian's chapter, reflecting the emergence into prominence of WikiLeaks as a new kind of non-state actor, just as this seminar got underway. This brings us to the important issues of uncovering occluded state data that are discussed by MacLean Hawley in his chapter on WikiLeaks.

Information

MacLean Hawley in his chapter, "Transparency Shift: An Overview of the Reality of WikiLeaks," makes a strong and convincing case of the changes wrought by the work of WikiLeaks and Julian Assange, especially during 2010 and 2011. Forms of resistance to the new imperial order, while not occupying an autonomous space within our seminar, are still considered to be central to our discussions. Hawley takes into account measures of public opinion, relies on primary source materials, and incorporates a range of experts' opinions (including by those who are conflicted by WikiLeaks), and defends an argument that few make, and that few have made so well: that WikiLeaks means empire cannot run the same way again. There is little doubt, after reading his chapter that WikiLeaks has become a prominent and critically important non-state actor in creating a new media landscape and helping to turn the tables on expanding state surveillance and the kind of secrecy apparatus built up in the name of national security.

Among the significant changes brought about by WikiLeaks, as identified by Hawley, we can list the following: 1) a change in the media-scape, that left the major international media players trying to catch up and mimic WikiLeaks; 2) the spawning of multiple, local leaks sites; 3) bringing the grotesque realities of war back into public view, and thus undercutting the increased spectacularization of American war (bitterly mocked by Mark Twain in Appendix C, "The War Prayer"); 4) contributing to increased public consciousness, and growing disapproval of the war in Afghanistan; 5) the creation of a free press haven in Iceland; 6) in Kenya, WikiLeaks' release of the Kroll report shifted a substantial mass of votes away from the leading presidential candidate, who allegedly had to rig the election in order to gain power; 7) much more debatable perhaps is the role of WikiLeaks in the Tunisian revolution in January 2011, although a number of writers credit a well-timed release of U.S. Embassy cables detailing extravagant corruption for inflaming the already heightened public animosity toward the regime; and, 8) "injured diplomacy," where U.S. influence and credibility were both diminished among some of its partners, some of whom, as in Afghanistan and Mexico, were flat out insulted and openly expressed their resentment and likely resistance to "business as usual" with the U.S.—this means a decline in U.S. soft power.

As anyone who has followed WikiLeaks would know, there is already a mountain of articles, essays, documentaries, and now books about the work of the organization. Having written articles about WikiLeaks myself, and trying to compile an ongoing bibliography of WikiLeaks-related news reports (and faltering, given the overload), I was especially impressed that Hawley undertook such an ambitious project, and delivered such a high-quality report as this one.

Ideologies: National Rhetoric; Humanitarianism

The chapter by Sabrina Guerrieri, "A War of Words: US-Iran relations in the Nuclear Debate," focuses on the role of national rhetoric in the mutual antagonism between Iran and the U.S., the work of rhetoric in shaping conceptions of the Other, influencing public opinion, and even working to reshape culture (the latter aspect not quite developed in this chapter). Guerrieri examines the role of the concept of sovereignty, and state rights, involved in producing national rhetoric and places it within the context of the conflict between the desire for energy independence, and imperial efforts to subordinate the potential military power of a rival.

In terms of how national rhetoric can shape public opinion at home, Guerrieri seems to be pointing out its successes. In shaping U.S. public opinion, American political leaders' rhetoric has most Americans perceiving Iran as potential nuclear threat. Here the phenomenon in question is that of demonization. In Iran instead there is a heightened rhetoric of victimization. Guerrieri outlines how Iran asserts its "indisputable nuclear right," condemning the imposition of "double-standards" by the UN Security Council, and the frequently stated resentment of discrimination by the West. Guerrieri goes one step further, articulating the position that Iranian national rhetoric is tied to controlling the domestic population, and in particular to diminish internal political opposition. To be fair, she could have applied the very same argument to the U.S.

Guerrieri concludes that there is a dangerous lack of trust produced between the two nations, and we can assume that it is one that helps to escalate the potential for direct conflict beyond rhetoric. She concludes that, "the 'bad guy' narrative of Iran projected by the U.S. must stop and the victimization narrative projected by Iran must in turn, begin to diminish". Looking at the anti-Iran and anti-U.S. rhetoric, which as public polls suggest are manufacturing public consent, Guerrieri believes this has the potential to stimulate detrimental outcomes, such as direct warfare. The role of ideology as expressed through rhetoric, where some values are taken for granted, and yet others are rendered explicit and defended vigorously, is a useful contribution to studies of the geopolitical conflict between the two states, for bringing attention to the importance of the role of underlying meaning in stimulating the conflict.

The problematic dimension to this analysis, in my view, is that it gives too much weight to rhetoric, as if the conflict between Iran and the U.S. was rooted in rhetoric, rather than the rhetoric rooted in their underlying conflict. It is not clear that a change in rhetoric will disturb the persistent realities of domination and resistance, and the differential structuring of power and legitimacy in the current world system. It is also my opinion that Guerrieri tended to downplay these considerations, and that she tended to err on the side of allegations that Iran is seeking, or even has, nuclear weapons—when in fact such allegations are themselves part of the production of demonizing rhetoric. Regardless, the diversity of perspectives on these issues is always welcome.

Industry: The Military-Industrial-Academic Complex in Canada and the U.S.

Laura Beach's chapter, "Canadian Academic Institutions, the Weapons Industry, and Militarist Ideology" explores the Canadian military-academic-industrial complex, presenting an overview of the involvement of Canadian universities in military research and the manifestations of military influence in the academic realm. Beach also considers some of the ethical implications of military research, with reference to the application of military technology developed by Canadian universities. In addition she addresses the question of academic objectivity of research on Canada's military that is funded by Canada's military, and how Canadian universities facilitate military research. Finally, she takes us to some of the university-based demilitarization campaigns.

This is a valuable contribution to a little researched area, especially where the production of knowledge in Canadian universities is concerned, and specifically the political economy of militarization. Indeed, even prior to publication in this volume, a shorter version of Beach's paper was widely circulated after it was published in *The Mark* and on the WikiLeaks news website, *WL Central*, as well as on the website of Anthropologists for Justice and Peace.

The scope and grasp of Beach's chapter are both considerable, covering: the advent of the military-academic-industrial complex and how it is has reshaped and redefined the role of scholars; the purposes of scholarship, and the orientations of knowledge production to serve the interests of empire, that is, of foreign intervention, the projection of military power, and the subordination of other nations; and, the alliance between the military, public institutions, and private corporations.

Beach begins by noting that "the corrupting power of the military-industrial complex was an overriding concern noted in President Dwight D. Eisenhower's farewell speech in 1961," which in itself bears a significance that may be taken for granted if not always understood: this is the voice of the ultimate insider, the commander-in-chief himself, not someone with clean hands, and as such plays the part of testimony, a confessional report of sorts, containing dire warnings from someone who was best placed to make them. For this reason it is reproduced in full as an appendix to this volume (see Appendix A). In this address, as noted by Beach, Eisenhower spoke of the "unwarranted influence" of the military industry on, as she puts it, "an increasingly militarized state" and which speaks "of the danger of domination by a militarized state on academic institutions". Eisenhower was of course not alone in understanding this, and not the first to write about it. Thus we include in Appendix B the entirety of Major General Smedley Butler's *War Is a Racket*, along with "War Is the Health of the State," by Randolph Bourne (Appendix D) and "The War and the Intellectuals," also by Randolph Bourne (Appendix E).

Beach, in adopting the concept of the "military-industrial-academic complex," underscores the influence of state, military and industrial interests on academic institutions. She demonstrates in her chapter that this "triangle of power" has been "amplified through a number of factors, including federal and state funding cuts to higher education, a growing emphasis on private-public research partnerships and the privatization of university research".

Beach's indictment of this reality is a strong one: "In conducting research and development that improves the capacity of the U.S. military, academic institutions promote the American imperialist project, the global 'war on terror' and the expansion of the American Empire". Her target is a formidable one, with over 350 colleges and universities in the U.S. (out of a total of 4,861) conducting Pentagon-funded research, so much so that the label, "militarized civilian university" seems appropriate. In Canada she counts 33 universities and colleges that conduct military-funded research—given that there are about 90 universities and colleges in all of Canada, the proportion here is striking. Indeed, one might argue that Canada has a greater spread of military-funded research than does the U.S.—almost 37% of Canadian universities and colleges, compared to 7% of U.S. universities and colleges that receive military research funding, albeit a superficial measurement that obscures many other factors.

Beach starts her analysis by examining the ways that the doors were opened to military-funded research in Canada, by first looking at the bases laid by shifts in research funding, and corporatization in the form of research that benefited private industry, including the weapons industries. Even public funding in many cases is tied to industry "matching funds". As she explains, "processes of corporatization open academia's doors to the influence of private military interests and war corporatism".

One of the persistent, though subtle themes in Beach's chapter has to do with impartiality and objectivity of academic work funded by the military and defence contractors. While she personally has been confronted by recipients of funding from the Security Defence Forum who argue that the Canadian military does not tell them what to say and does not stymie what they choose to publish, there are larger issues that SDF supporters are ignoring. First is the issue of self-selection: those who are amenable to defence funding, and who see their work fitting within the parameters of research supported by SDF,[1] are the ones to first select themselves as possible candidates for military funding because of the basic amenability of their research projects. Second is the issue of what we might call structural partiality—as impartial as these scholars believe their thinking to be, the fact is that by seeking and accepting such funding they reinforce and legitimate the military presence in academic research. This reinforces the structural power of the military branch of the state, by accepting it as a legitimate arbiter of research funding, as an authority on what kind of academic thinking should be rewarded. Third is the issue of commonplace, taken for granted silencing: while there is a SDF to fund research that the military finds amenable to its interests, there is absolutely no institution that funds anti-war research, peace research, or endows chairs in anti-imperialist studies. This is quite astounding, especially when we consider that for a decade the majority of Canadians have consistently opposed military intervention in Afghanistan—yet their universities, supposedly *public*, are being harnessed to serve aims inconsistent with the public interest, hijacked by both the state and narrow private sector interests. It is thus interesting to note, as Beach does, that two out of three "representatives" of the "community at large" that sit on the board of the SDF, are private defence contractors. This type of distortion and misappropriation of "community at large," used as a fig leaf for corporatization, is now increasingly common in Canadian universities. Indeed, in the case of Concordia University, to which Beach devotes attention, five of the members of the Board of Governors, so-called representatives of the community at large, are themselves defence contractors.

At any rate, the question of whether or not militarist ideology is perpetuated via military funding is only raised when it applies to social science research. At the same time, a great deal of funding is directed to the development of weapons systems by various engineering departments across Canada—and here there can be little room for doubt about the goals served by academics engaged in such research. The ideology here is given, working at its ideological best by removing itself from question, or even discussion.

Where there is ample and justified room for concern is in the clear pattern of dismissing ethical issues in the use of university research for the purpose of killing other human beings. In the case of Concordia University, a special focus of Beach's research, she found that when the Faculty of Computer Science and Engineering was asked about "its relationship with Pratt and Whitney in light of the company's involvement in supplying the Indonesian military with tools of genocide used in East Timor," Dean Nabil Esmail answered only that the faculty had a "longstanding and constructive relationship" with the company for decades. Indeed, representatives of the company have sat on the Faculty's own External Advisory Boards, thus further erasing the line between the university, private industry, and the military. Moreover, as Beach explains, "there are currently no policies regarding the ethical implications of military research outside of the university sphere at any university in Canada," and hence little consideration, beyond specific research procedures, of how research funded by the military, and for the purpose of developing or contributing to weapons systems, is a fundamental violation of the Canadian Association of University Teachers' stated principle that "education's most basic purpose is to enhance life and the dignity of the human person an objective that is difficult to achieve in the absence of fundamental human rights". When it comes to military research, only then are codes of ethics considered a violation of academic freedom—for war, there is to be absolute and total immunity and impunity for research, that is, when research is most clearly designed to do harm.

Notes

1 See the Security and Defence Forum: Academic Programs, at http://www.forces.gc.ca/admpol/SDF-eng.html

References

Chomsky, N. (1991). International Terrorism: Image and Reality. *Chomsky.info*, December.
    http://www.chomsky.info/articles/199112-02.htm
———— . (2002a). Noam Chomsky on the Middle East and the US War on Terrorism. *Dissident Voice*, July 28.
    http://dissidentvoice.org/Articles/Chomsky_DV-HotType.htm
———— . (2002b). Who are the Global Terrorists? *Chomsky.info*, May.
    http://www.chomsky.info/articles/200205-02.htm
———— . (2002c). Terror and Just Response. *Chomsky.info*, July.
    http://www.chomsky.info/articles/20020702.htm
Ignatieff, M. (2004). *The Lesser Evil: Political Ethics in an Age of Terror*. Toronto, ON: Penguin Group (Canada).
Lock, E. (2010.) Soft Power and Strategy: Developing a "Strategic" Concept of Power. In I. Parmar & M. Cox (Eds.), *Soft Power and US Foreign Policy: Theoretical, Historical and Contemporary Perspectives* (pp. 32-50). New York: Routledge.
Siddiqui, N. (2010). Torture and the Global War on Terror. In M.C. Forte (Ed.), *The New Imperialism, Volume 1: Militarism, Humanism, and Occupation* (pp. 111-123). Montreal, QC: Alert Press.

CHAPTER ONE

CANADIAN ACADEMIC INSTITUTIONS, THE WEAPONS INDUSTRY, AND MILITARIST IDEOLOGY

Laura Beach

"In the councils of government, we must guard against the acquisition of unwarranted influence, whether sought or unsought, by the military industrial complex. The potential for the disastrous rise of misplaced power exists and will persist....The prospect of domination of the nation's scholars by Federal employment, project allocations, and the power of money is ever present and is gravely to be regarded." (Eisenhower, 1961, pp. 1038-1039)

The corrupting power of the military-industrial complex was an overriding concern noted in President Dwight D. Eisenhower's farewell speech in 1961. In this address Eisenhower spoke of the "unwarranted influence" of the military industry on an increasingly militarized state and of the danger of domination by a militarized state on academic institutions. The link between academia and the military-industrial complex was still a predominant issue less than a decade later. Senator J. William Fulbright (1970) warned strongly against the use of the university as an instrument for government and military purposes, coining the term "military-industrial-academic complex". The influence of state, military and industrial interests on academic institutions has been amplified through a number of factors, including federal and state funding cuts to higher education, a growing emphasis on private-public research partnerships and the privatization of university research (Bernans, 2001; Giroux, 2009). Together this triangle of power effectively redefines "the nature of research, the role of faculty, the structure of university governance, and the type of education offered to students" (Giroux, 2009, ¶ 7). Gone is the role of the university as the bastion of civic, critical and democratic education, here is the university as an extension of state ideology, interests and power, or as Henry Giroux (2007) has so aptly coined, the "university in chains".

The militarization of academic institutions is a crucial part of the new American imperialist project, which "envisages a coercive refashioning of the world to suit American interests" (Forte, 2010, p. 5). America's "informal empire" seeks to project U.S. military power and American ideals to every corner of the world, through economic integration and unilateral military supremacy (Johnson, 2000; Forte, 2010). In conducting research and development that improves the capacity of the U.S. military, academic institutions promote the American imperialist project, the global "war on terror" and the expansion of the American Empire.

As Canada undertakes a similar transition from welfare state to warfare state, the same processes of corporatization and militarization are hard at work re-designing the focus and purpose of Canadian academia. Numerous Canadian universities develop technology towards the creation of weapons of war, engage in social and political science research funded by the Canadian military and in so doing contribute to the expansion of military presence, armament and intervention, American and otherwise, across the globe.

This chapter explores the Canadian military-academic-industrial complex, presenting an overview of the involvement of Canadian universities in military research and the manifestations of military influence in the academic realm. The ethical implications of military research will be discussed with regards to the application of military technology developed in part by Canadian universities, the question of academic objectivity and critical capacity of research on Canada's military that is funded by Canada's military, and the role that Canadian universities play through these means in facilitating military endeavours. The position of the opposition movement which calls for the demilitarization of the university will also be presented here, including specific campaigns targeting military research conducted by Canadian professors and graduate students.

The Military-Academic-Industrial Complex

The military-academic-industrial complex has two distinct arms: the overt blend of military and academia embodied by military colleges and universities and the more covert influence of military interests manifest in what Nick Turse refers to as the "militarized civilian university" (Turse, 2004, ¶ 2). In addition to military-run educational institutions, including the numerous schools of the National Defense University system, in the U.S. today there are approximately 350 colleges and universities conducting Pentagon-funded research (Turse, 2004, ¶ 10). The Department of Defense is the third largest federal funding source for university research (Turse, 2004, ¶ 10).

In Canada there are a number of academic institutions overtly dedicated to military purposes, including the Royal Military College of Canada in Kingston, Ontario, the Royal Roads Military College in Victoria, British Columbia and the Royal Military College in Saint-Jean, Québec. The number of "militarized civilian universities" is considerably greater. Dalhousie University, Université Laval, Memorial University, University of Toronto, Université de Montréal, Manitoba University, Western University, Calgary University, Sherbrooke University, Queens University, Institut National de la Recherche Scientifique, Carleton University, University of Alberta, University of Regina, York University, Ottawa University, Ryerson University, University of British Columbia, McGill University, McMaster University, Université du Québec à Montréal (UQAM), Simon Fraser University, University of Saskatchewan, University of New Brunswick, University of Waterloo, Moncton University, Saint Mary's University, University of Victoria, Concordia University and Windsor University all receive funding contracts from the Canadian Department of National Defence (Public Works and Government Services Canada, 2011).

The expansion of the military-academic-industrial complex in Canada can be linked to the same factors at work in the United States: federal and provincial funding cuts to higher education, a growing emphasis on private-public research partnerships and the privatization of university research.

Performance-Based Funding, Public-Private Partnerships and Proprietary Research

Since the 1970s North American academic institutions have experienced a radical shift in criteria for federal and state/provincial funding. Whereas funding was once tied to factors such as enrolment, retention and graduation rates, shifts in the funding paradigm have introduced new "performance" indicators including faculty workload and productivity, job placement data and external (corporate) research funding (Bernans, 2001). As David Bernans (2001, p. 20) has noted, "'performance' has universally been interpreted as increasing corporate satisfaction with the 'products' (students and research) measured in terms of students' success on the labour market or the production of commercially viable knowledge, and the efficient production of those products".

At the same time as shifts in the funding paradigm were taking place, the United States Business Roundtable and the Canadian Business Council on National Issues were hard at work lobbying towards policies encouraging publicly-funded research institutions to patent and sell findings and inventions to private industry (Bernans, 2001). Prior to these policies the nature of university research had been non-proprietary, published in a public domain and free to use or exploit. New sales and revenues from royalties motivated public institutions to keep their research under wraps; simultaneously corporations were given generous tax incentives to fund university research (Bernans, 2001). Thus was born the public-private research partnership. Much of the funding to academic institutions from the Canadian government is now tied to industry funding through "matching funds" projects. For example, the Canadian Foundation for Innovation gives matching funds to projects that bring profits to private industry while the Natural Sciences and Engineering Research Council of Canada (NSERC) and the Networks of Centres of Excellence (NCE) favour research projects with direct applications in private industry (Tudiver, 1999). The Industrial and Regional Benefits (IRB) Policy, created in 1986, "provides the framework for using federal defence procurement to lever long-term industrial and regional development within Canada" (Industry Canada, 2011, ¶ 2). Though not specifically geared towards university research partnerships, many universities have been benefiting from the IRB policy through which matching funds for research and development are provided by industry contractors and Regional Development Agencies.

Funding for non-proprietary research has become a thing of the past. Intellectual proprietorship policy changes, tax breaks and shifts within the federal funding paradigm have amplified the influence of industrial corporations over research and curriculum within Canadian academic institutions. The changing role of the university, from a place of critical thinking, relatively unaffected by the demands of industry and the state to an instrument of industry and state interests, is reflected in statements by industry professionals. The Chairman for Whirlpool Corporation, for example, has remarked that "we could not work with a university that has a long drawn-out process for working with corporations. Rather, universities must listen to the needs of corporate customers and then quickly marshal the resources to solve the learning requirement" (Whitwam quoted in Meister, 1998, p. 188). Indeed, critics of the corporatization of the Canadian university have argued that today "it is the corporation that identifies the learning requirement" and by extension the kind of curriculum that is needed. The role of the university is simply to "marshal the resources" necessary to fulfil the task" (Bernans, 2001, p. 47). Processes of corporatization open academia's doors to the influence of private military interests and war corporatism.

Manifestations of Militarization

The militarization of university research and curricula in Canada takes many forms. The Department of National Defence funds social and political science research, producing "impartial" academic opinions on the state of Canada's military. Critics claim that the resulting publications encourage the public to accept the expansion of military presence, armament and intervention, aiding in the justification of state military spending and military intervention (Operation Objection, 2008). Supporters claim that DND does not impose any paradigm or dictate any research outcomes, although they bypass the question of self-selection (those who wish to produce work that is amenable to agencies such as the DND, seek out its funding). Funding for the development of military technology within university departments of Science and Engineering comes from DND, Defence Research and Development Canada (DRDC), NSERC, the National Research Council (NRC), Precarn and the Intelligent Sensing for Innovative Structures (ISIS) Canada Research Network as well as through various branches of the U.S. military, including the Defence Threat Reduction Agency (Operation Objection, 2008). While these lines of influence are relatively easy to document, funding also comes in the form of corporate research partnerships and donation agreements which are not held to the same requirements of transparency as federal funds. Corporations sometimes sub-contract universities to complete work commissioned by them from the federal government. In playing the middle-man, industry acts to obscure processes of militarization which are nevertheless in action (Operation Objection, 2008).

The Security Defence Forum (SDF)

The SDF was established by the Department of National Defence (DND) during the Cold War with the mandate to redistribute military funds to "centres of expertise" within Canadian political science departments. These research centres specialize in military-related fields, analysing the Canadian army's actions and priorities and presenting their findings as "impartial" experts to various media outlets (Operation

Objection, 2008). While the SDF is not the sole funding source, they do provide a significant amount of funding for these centres, representing a direct line of influence from the Canadian military to political science professors and students within Canadian Universities. The DND claims that "academic freedom and objectivity are at the heart of the Forum's mandate" (National Defence and the Canadian Forces, 2010). However, demilitarization groups such as Operation Objection are quick to point out the impossibility of impartiality that ties between the SDF and the military and weapons industries present. For example, during the fiscal year 2007-2008 *two out of three representatives of the community at large* on the SDF selection committee, the body in charge of allocating SDF resources to "centres of expertise," had direct links with the military-industrial complex (Security Defence Forum [SDF], 2008). Mr. Arthur C. Perron is the retired Vice–President Communications at CAE, a high-grossing military weapons systems provider that was recently awarded military defence contracts with over 12 countries collectively valued at more than $140 million (SDF, 2008; Defense World, 2011). H. Cameron Ross is a retired Military-General and the Senior Military Advisor to EnCana, one of the world's largest natural gas companies, grossing U.S.$4.4 billion in 2010 (SDF, 2008; EnCana, 2011). Having representatives from corporations with major economic interests in military actions, operations and interventions on a committee that effectively decides the nature of research to be funded seriously cast doubts on the "objective" nature of research undertaken in these centres. Research topics funded by the SDF include: terrorism; weapons of mass destruction; Canadian forces transformation; Canada-United States defence relations; the Canadian Forces' international role and Defence procurement and management.

Through the SDF, tax payers' dollars are funnelled through the federal government and the Canadian military to subsidize the salaries and research expenses of professors and graduate students from 14 Canadian universities operating a total of 12 "centres of expertise" (SDF, 2008). These universities include: Concordia University, Dalhousie University, University of New Brunswick, Queen's University, Carleton University, York University, Wilfrid Laurier University, University of Manitoba, University of Calgary, University of British Columbia, Université de Montréal, McGill University, Université du Québec à Montréal and Université Laval. During the 2007-2008 fiscal year these "centres of expertise" received $1,620,000 in SDF funding (see Table 1), $392,500 in Academic Awards, $268,963 in Special Projects Funds, $89,385 in International Conference Funds and $32,151 in National Conference Funds for a grand total of $2,402,999 (SDF, 2008).

Table 1.1: Security Defence Forum Funding to "Centres of Expertise" 2007-2008 (Source: SDF, 2008)

Centre of Expertise	University Partners	Amount of Funding
School of Policy Studies (Chair of Defence Management Studies)	Queen's University	$165, 000.00 (salary for Chair of Defence Management Studies)
Centre for Security and Defence Studies	Carleton University	$140,000.00
Centre for Military and Strategic Studies	University of Calgary	$140,000.00
Centre for Foreign Policy Studies	Dalhousie University	$140,000.00
Centre of Defence and Security Studies	University of Manitoba	$120,000.00
Centre of International Relations	University of British Columbia	$120,000.00
Gregg Centre for the Study of War and Society	University of New Brunswick	$120,000.00
Research Group in International Security	McGill University Université de Montréal	$120,000.00
Centre for Military, Strategic and Disarmament Studies	Wilfrid Laurier University	$115,000.00
Centre for International Relations	Queen's University	$115,000.00
Programme Paix et Sécurité Internationale	Université Laval	$115,000.00
Centre d'Études des Politiques Étrangères et de Securité	Concordia University Université du Québec à Montréal	$110,000.00
Centre for International and Security Studies	York University	$100,000.00
Total amount of funding		$1,620,000.00

Over the same fiscal year, the SDF centres and the SDF Chair shared their perspectives on Canadian militarism through events, media interviews and Op-ed articles. They collectively conducted over 1,300 media interviews, wrote approximately 100 Op-ed articles and hosted 412 events that reached more than 18,000 people (SDF, 2008). Operation Objection (2008) has pointed to a correlation between increased funding to "centres of expertise" and escalating aggression in military interventions, further highlighting the role that SDF funded centres play in the justification of military exploits. In 2006 Canadian troops in Afghanistan moved south to Kandahar, an area with considerably more instability and combat. Predictably, the number of casualties of war significantly increased, calling into question the nature of Canada's "peace-keeping" mission. Over the 2005-2006 fiscal year the total SDF budget had increased by 25% (Operation Objection, 2008).

SDF funded professors and graduate students are in no way merely instruments for the Canadian military with no critical capacity and individual ideology. However, the SDF program does provide concrete evidence of the influence and prevalence of the Canadian military in the academic funding paradigm. It is also important to note that there is no equivalent federal funding body for peace research; this disparity highlights the pervasive nature of the military-industrial complex in Canadian academia.

Research and Development of Military Technology

In addition to the perpetuation of militarism through government funded social and political science research there is an incredible amount of federal funding directed towards the development of military technology. However, only contracts deemed "unclassified" are issued publicly, and even then the exact details of these contracts are difficult to obtain. As stated above, corporate funding of military technology is also not required to be issued publicly; it is therefore impossible to ascertain exactly how much corporate money flows into universities for the purposes of military technology development. This makes it impossible to know the full scale of funding towards military technology within Canadian universities.

Department of National Defence and Defence Research and Development Canada

DND is one of the main sources of federal funding towards military technology for Canadian universities. During the fiscal year 2010 to

2011 DND issued a total of $17,371,606 in contracts to Canadian Universities (Public Works and Government Services Canada, 2011). DRDC is also a significant funding source, dedicating roughly half of its approximately $300,000,000 annual budget to collaborative programs with private industry and the academic community (Defence Research and Development Canada [DRDC], 2010). DRDC has three research partnership programs aimed at encouraging collaboration between DND, private industry and academic institutions: the Defence Industrial Research Program (DIRP), the Technology Demonstration Program (TDP) and the DND/NSERC Research Partnership Program (DRDC, 2010). The DND/NSERC Research Partnership Program aims to build "strong three-way linkages and create synergy between researchers in DND and universities and the private sector" (DRDC, 2010). Recent research partnerships funded in part by the DRDC include SMARRT and JSMART Techologies.

SMARRT Technology

SMARRT stands for Simulation and Modeling for Acquisition, Requirements, Rehearsal and Training and represents a new partnership between the federal government, industry aimed at establishing Canada as a world leader in Modeling and Simulation (Vallerand et al., 2005). SMARRT technologies are being used to develop and improve the Canadian Forces' military systems and capabilities. The October 2004 UAV-NTS SE Interoperability Experiment was the first step towards SMARRT partnerships between DRDC, the military industry and Canadian Universities, linking DRDC Ottawa with Carleton University in developing Uninhabited Air Vehicle (UAV) research. Research sessions were attended by prospective government, academic and industry partners and foreign military personnel (Vallerand et al., 2005).

JSMARTS Technology

JSMARTS stands alternatively for Joint Simulation, Modeling for Material Acquisition, Requirements, Training and Support or Joint SMARRT. This program minimizes the risk associated with networking UAV technology partners by linking partners through a Wide Area Network (WAN) distributed simulation. Through this network military agencies including DRDC Ottawa and NATO, military industry corporations such as CAE Inc. and Canadian academics can interface safely towards the development of UAV technology: "while CAE Inc. was providing CGF (Computer Generated Forces), DRDC Ottawa's FFSE Section was providing a UAV (Uninhabited Aerial Vehicle) simulator, and a NATO STANAG 4586 compliant Ground Control Station, both supported by the Joint Simulation Network" (Kim et al., 2005, ¶ 1).

Unmanned Vehicle Technology

The most common military weapons currently being developed within Canadian universities are Unmanned Aerial Vehicles (UAV). In addition to universities specifically developing UAV technology, including Concordia University, University of British Columbia, Simon Fraser University and Université de Sherbrooke, there are a number of universities that are currently developing technology directly applicable to Unmanned Vehicles (UV). DRDC has compiled information on all Canadian universities undertaking such technology in a report entitled "Technological Drivers and Constraints related to Autonomous Collaborative Unmanned Vehicles" funded by the Canadian Forces Experimentation Centre (CFEC) (Bowen & MacKenzie, 2003). The expressed goal is to develop Autonomous Collaborative UVs, including Unmanned Aerial Vehicles (UAV), Unmanned Cyber Vehicles (UCV), Unmanned Ground Vehicles (UGV), Unmanned Orbital Vehicles (UOV), Unmanned Surface Vehicles (USV) and Unmanned Underwater Vehicles (UUV), for use in Department of National Defence missions by the year 2025: "these unmanned vehicles, or robots, would support the execution of dull, dirty and dangerous missions and will reduce the likelihood of military personnel being placed in harm's way" (Bowen & MacKenzie, 2003, p. 1). In this report DRDC expresses the intent to utilize existing agreements between DND and NSERC, PRECARN/IRIS and NRC for partial funding of DND programs to ensure that research funded by these bodies supports DND's goals. The tables below were adapted from tables provided in the DRDC report (Bowen & MacKenzie, 2003). Table 2 showcases the amount of funding allocated to specific universities involved in UV applicable technology in the form of NSERC research grants and scholarships. Table 3 provides an overview of the type of technologies specific universities are developing that either currently contribute, or could potentially contribute to the development of UVs. During the year 2002 NSERC alone provided $6,784,318 in research grants and scholarships to professors and graduate students working on technology applicable to UVs (Bowen & MacKenzie, 2003).

Table 1.2: NSERC/DRDC Funding to Canadian Universities towards Enabling Technology for Autonomous Collaborative Unmanned Vehicles (2001-2002).

Enabling Technology	Canadian Research Effort # of Grants Average $/Grant	Highest Intensity of Effort (University Locations)	$ Currently being Expended
Intelligent S/W, Sensing and Navigation	108 $29,000/grant	Toronto, McGill, Alberta, Ottawa, Windsor, UBC	$3,107,000.00
Control Systems	32 $41,000/grant	UBC, Western	$1,320,000.00
Robotics, Dynamics and Mobility	29 $32,000/grant	Laval, École Poly, Carleton, Queen's, Guelph, Ottawa, Waterloo	$936,000.00
Materials	12 $35,000/grant	McGill, Sherbrooke, Alberta, Laval, Queen's, UBC, Manitoba, Waterloo	$415,000.00
Power/Energy Systems	7 $39,000/grant	Concordia, Dalhousie, McGill,	$274,000.00
Communications (see note below)	2 $29,000/grant	Simon Fraser, Alberta	$58,000.00
Total amount of funding			$6,110,000.00

Note: Communications funding was heavily influenced by private-sector activity.
Source: Bowen & MacKenzie, 2003

Table 1.3: Research Universities and Enabling Technology Areas

University	Robotics	Dynamics	Mobility	Navigation	Intelligence Software	Sensing	Communications	Control Systems	Power	Materials
Brock University					X					
Carleton University		X			X		X	X		
Concordia University		X				X		X	X	
Dalhousie University			X		X	X		X		
McGill University		X			X	X		X	X	X
McMaster University		X			X	X				
Memorial University	X		X		X					
Queen's University		X		X	X	X		X		X
Royal Military College of Canada	X				X					X
Ryerson University					X	X		X		
Simon Fraser University		X			X	X	X	X		
University of Alberta					X	X	X	X		X
University of British Columbia		X			X	X		X		X
University of Calgary		X			X	X		X		
University of Guelph		X			X	X				X
University of Manitoba		X			X	X		X		X
Université de Montréal		X			X	X				
University of New Brunswick					X			X		
University of Ottawa		X			X	X				X
Université du Québec		X			X	X		X		X
University of Regina					X					

University of Saskatchewan				X	X	X			
University of Toronto				X	X	X	X		
University of Victoria				X					
University of Waterloo	X			X	X	X		X	
University of Western Ontario	X			X	X	X			
University of Windsor				X	X				
York University	X			X	X				

Source: Bowen & MacKenzie, 2003.

Additional Contributions by University

This section details some specific contributions of Canadian universities towards the development of weapons of war, in addition to what has been outlined above. It is by no means a comprehensive overview.

University of British Columbia and Simon Fraser University

The University of British Columbia houses Thunderbird Robotics, a group dedicated to researching and developing autonomous robots. Formed in 2004 by Professor John Meech of the Norman B. Keevil Institute of Mining Engineering, Thunderbird Robotics has competed in the 2005 DARPA Grand Challenge and the 2007 DARPA Urban Challenge, organized by the Defence Advanced Threats Research Projects Agency of the U.S. military (Centre for Environmental Research in Minerals, Metals and Materials [CERM3], 2011; UBC Thunderbird Robotics, 2011; Tether, 2006). Over 250 undergraduate students have been involved in this research (CERM3, 2011).

Professor Vikram Krishnamurthy and Michael Maskery from the UBC Department of Electrical and Computer Engineering have received funding from the DND and worked with DRDC employee Christina O'Regan towards a game-theory model for naval missile defence (Maskery, Krishnamurthy, & O'Regan, 2005).

In 2010 researchers from Simon Fraser University and the University of British Columbia joined forces with the Boeing Company in the Vancouver Institute through the IRB program in the creation of the Vancouver Institute for Visual Analytics (VIVA, 2011). Through the Boeing Visual Analytics Research Grant the Canadian Forces acquired four C-17 Globemaster III multi-mission transport aircraft and fifteen CH147 medium-to-heavy-lift helicopters (VIVA, 2011).

Carleton University

Professors Jo-Anne LeFevre and Chris Herdman of the Centre for Applied Cognition in Carleton's Department of Psychology have been awarded a DRDC contract for simulations of pilot awareness systems in military aircraft. Herdman has a history of research partnerships with DRDC, having worked with them towards the development of a simulator for pilot-awareness for military helicopters (Operation Objection, 2008).

Memorial University

Investments from the Boeing Company and the Research and Development Corporation of Newfoundland and Labrador through the IRB program have helped to create a Mechatronic Development and Prototyping Facility at Memorial University of Newfoundland. Dr. Nicholas Krouglicof of Memorial's Faculty of Engineering and Applied Science has emphasized the role of the facility in the development of sensor technology and intelligence systems to improve the autonomy and reliability of UAVs (Locke & Whelan, 2011).

Université de Sherbrooke

At Université de Sherbrooke a group of undergraduate students in the Department of Mechanical Engineering formed Véhicule Aérien Miniature de l'Université de Sherbrooke (VAMUdeS), a research group dedicated to the development of UAV technology. The university provided funding and resources for VAMUdeS, of which many members are now employed by military subcontractors (Operation Objection, 2008). VAMUdeS successfully competed in the 2006 Forces Avenir Contest and won first place in the UVS Canada UAV Competition 2006-2007 (VAMUdeS, 2007). The group's corporate sponsors include Pratt & Whitney Canada, Lyrfac AAI and Groupe CMI (VAMUdeS, 2007).

McGill

The McGill Department of Mechanical Engineering has a long history with the development of warfare technology. Since the 1960s the department's Shock Wave Physics Group (SWPG), formerly known as the Gas Dynamics Laboratory, has been involved in research towards Fuel

Air Explosives (FAE) funded first by the U.S. Air Force Office of Scientific Research (AFOSR) and the U.S. Defense Threat Reduction Agencies (DTRA) and then by DRDC through the Defense Research Establishment Suffield to Investigate Fuel-Air Explosives (DRES) (Higgins, 2009). Professors and graduate students have worked closely with personnel from private military corporations, including Combustion Dynamics, and federal military agencies such as DRES and DRDC. There has also been cross-seeding between various institutions, with McGill students going on to work for DRES or private military corporations and DRES employees attending McGill for graduate studies and conducting research with the SWPG.

Early FAEs developed in part by professors at McGill were used in the Vietnam War in the 1970s. FAEs release a "high pressure shock wave that expands with a force strong enough to flatten trees or buildings" (Higgins, 2009, p. 6). Persons nearby are crushed, incinerated or asphyxiated, those further away experience internal rupturing; the effect has been compared to that of a small nuclear bomb (Higgins, 2009). FAE bombs were also employed during the 1991 Gulf War in Iraq and Kuwait. New development of this technology, renamed "thermobaric explosives", or Solid Fuel Air Explosives (SFAE), was prompted by the United States military's desire to kill people hiding in tunnel complexes in their invasion of Afghanistan in 2001 (Higgins, 2009). This technology was used for the first time in targeting a cave complex in Afghanistan in 2002. It is interesting to note that the McGill research and development that contributed towards SFAE technology took place after funding from AFOSR stopped and the SWPG was funded only through the Canadian military. This demonstrates that technology doesn't have to be funded by the U.S. military to be useful to and used by the U.S. military (Higgins, 2009).

The Case of Concordia University

Background

When Loyola College and Sir George Williams University fused together to form Concordia University in 1974, Concordia's mission statement emphasized the commitment of the university to "responsible and innovative leadership in fulfilling the mission of universities to develop and disseminate knowledge and values and to act as a social critic" and to preparing graduates to live as "informed and responsibly critical citizens" (Concordia's Thursday Report, 2001). The name Concordia comes from the motto of the City of Montréal—Concordia Salus—which means "well-being through harmony" (Concordia University, 2011b). However, the processes of corporatization and militarization described above have been no less pervasive at Concordia than elsewhere. The shift to performance-based federal funding was echoed by the Québec provincial government, with an emphasis on "market-based" funding criteria including public-private partnerships and corporate interest (Bernans, 2001). Concordia's links to the military-industrial complex are numerous, calling into question whether the university continues to abide by its original mission statement or the connotations of its own name.

Centre d'Étude des Politiques Étrangères et de Securité (CEPES)

Concordia University is a partner with UQAM in CEPES, a SDF funded "centre of excellence" that receives on average annual funding of $110,000. Concordia professors involved in this centre include: Julian Schofield, Michael Lipson, Norrin Ripsman, Peter Stoett and Elizabeth Bloodgood (Centre d'Étude des Politiques Étrangères et de Securité [CEPES], 2011). Their most recent CEPES publications are listed below (CEPES, 2011).

Julian Schofield:
★ Islam and the Afghanistan Campaign (2010)
★ Appraising the Threat of Islamist Take-Over in Pakistan (2007)
★ Challenging Conventional Wisdom—Debunking the Myths and Exposing the Risks of Arms Exports Reform (2004)
★ Arms Races and War in the Indo-Pakistan Rivalry, 1947-1971
★ Special Operations and Unconventional Warfare (2003)
★ Kidnapping and Extortion: Common Terrorist Tactics (2003)

Michael Lipson:
★ Transaction Cost Estimation and International Regimes (2004)

Norrin Ripsman:
★ Assessing the Uneven Impact of Global Social Forces On the National Security State: A Framework for Analysis (2005)
★ The Politics of Deception: Forging Peace Treaties in the Face of Domestic Opposition (2004)
★ Under Pressure: Globalization and the National Security State (2004)

★ Qualitative Research on Economic Interdependence and Conflict: Overcoming Methodological Hurdles (2003)

Peter Stoett:

★ Bilateral Ecopolitics: Continuity and Change in Canadian-American Environmental Relations (2006)
★ International Ecopolitical Theory: Critical Reflections (2006)
★ Global Politics: Origins, Currents, Directions (2004)

Elizabeth Bloodgood:

★ Elizabeth has no CEPES publications listed on the CEPES website.

Department of National Defence/RCMP Funding

Concordia also receives federal military funding through contracts with the Department of National Defence and the Royal Canadian Mounted Police. While the date, amount of funding and funding contract numbers are easily accessible online, contract details are not made available.

Table 1.4: DND/RCMP Funding to Concordia University, 2008-2011

Contract Number	Award Value	Award Date	Classification	Client
M9010-084422/ 001/SS - 000	$880,010.00	03/25/08	Information Retrieval Services, Database	RCMP
W7701-081745/ 001/QCV - 000	$141,750.00	12/03/08	Military	DND
W7701-101423/ 001/QCL - 000	$24,000.00	07/08/10	Military	DND
W7701-103524/ 001/QCN - 000	$24,457.00	02/07/11	Informatics Professional and Consulting Services, Systems Analysis Design	DND

Source: Public Works and Government Services Canada, 2011.

Pratt and Whitney and the Concordia Institute of Aerospace Design and Innovation

Concordia's Faculty of Computer Science and Engineering (CSE) has been extremely successful in garnering corporate funding towards military technology. This is at least in part due to the lack of consideration of the ethical implications of said funding on the part of faculty administration (Bernans, 2001). When asked about CSE's relationship with Pratt and Whitney in light of the company's involvement in supplying the Indonesian military with tools of genocide used in East Timor, CSE Dean Nabil Esmail answered only that the faculty has had a "long-standing and constructive relationship" with the company for decades (Concordia Senate, 1999). Pratt and Whitney has also been linked to the production of F100-PW-229 engines for the Israeli military's F-16 fighter jets. These jets have been employed in the bombing of Palestinian civilians; however this has not deterred the Concordia administration and the CSE from doing business with Pratt and Whitney. The company has given CSE hundreds of thousands of dollars in research contracts and donations and has sat on the Faculty's External Advisory Board (Bernans, 2001). Between 1996 and 1999 Concordia was awarded $500,000 in research contracts from Pratt and Whitney (Concordia University, 1999). In 1998 the company gave $100,000 towards the construction of an Aerospace Simulation Facility and in 2000 they provided an additional $1.2 million towards the creation of the Concordia Institute of Aerospace Design and Innovation (CIADI) (Lacome, 1998; Vincelli, 2000). Through the CIADI Concordia professors and students are currently involved in research projects with Pratt and Whitney, Bombardier Aerospace, Aviya Technologies, Bell Helicopter Textron and Thales, all corporations specializing in technology employed in military operations (Concordia Institute of Aerospace Design and Innovation [CIADI], 2011; Bombardier, 2011; Aviya, 2008; Bell Helicopter Textron, 2011; Thales, 2011). Bell Helicopter and Thales both contribute to warfare technology used by the U.S. military (Bell Helicopter Textron, 2011; Thales, 2011).

Unmanned Aerial Vehicles

Concordia has two labs dedicated to UAV research and development: the Diagnosis, Flight Control and Simulation (DFCS) Lab; and, the Networked Autonomous Vehicles (NAV) Lab. Both labs are connected to the Department of Mechanical and Industrial Engineering and CIADI. Through these labs Concordia professors, Ph.D. students, Masters students and Undergraduate students conduct research and develop UAV technology. Since 2007 over 30 students have been involved in the UAV labs (Concordia University, 2011a). Current research projects include collaborative research with DRDC on "Fault Diagnosis, Fault Tolerant Control and Cooperative Control" of UAVs and collaborative research with DRDC, Quanser Inc., Opal-RT Technologies

Inc., and Numerica Inc. on "Cooperative UAV Systems with Increased Reliability in Severe Environments" (Concordia University, 2011a). The latter research project is funded through a NSERC Strategic Project Grant. Professor Youmin Zhang of the Department of Mechanical Engineering has been instrumental in much of the UAV research undertaken in the DFCS and NAV labs at Concordia. Since 2005 Zhang has presented at 28 Research Seminars and Conferences at universities in Canada, China, Poland and France and at a DRDC facility in Valcartier, Québec (Concordia University, 2011a). Zhang is currently co-coordinating the Third Symposium on Small Unmanned Aerial Vehicle Technologies and Applications (SUAVTA), to take place 29-31 August 2011 in Washington DC (Concordia University, 2011a).

Concordia's Board of Governors

In addition to the CEPES partnership and funding from the DND, RCMP and military corporations towards the development of military technology, Concordia University has ties to the military-industrial complex through the members of its Board of Governors (Concordia University Board of Governors [BoG], 2011). Concordia's BoG is the highest decision-making body within the University. The power of the Board includes the establishment of legal and administrative frameworks for the University as well as "superintending and reforming power over all decisions affecting activities held at the University or connected with the University" (BoG, 2011). Of the 40 voting members of the BoG, 23 are representatives from the "community-at-large". The vast majority of these representatives are Presidents, Directors, Chief Executive Officers (CEO) or Partners of multi-million dollar corporations. A few members have direct links to the military-industrial complex.

Peter Kruyt
Chair of the BoG, Peter Kruyt is the President and CEO of Victoria Square Ventures, Inc. He is also the Director of China International Trust and Investment Corporation (CITIC) Pacific—a multifaceted corporation involved in iron ore mining, steel manufacturing, property development and investment, power generation, aviation, tunnels, telecommunications and motor vehicle and consumer product distribution (CITIC Pacific, 2011). CITIC Pacific is the parent company of Poly Technologies (Baoli) Ltd., a defence manufacturing company specializing in medium and long-range missiles that was once one of China's two major arms trading organizations (NTI, 1998). Poly Technologies is

responsible for the illegal shipment of 2,000 AK-47s to the U.S. in 1997 (Medeiros & Gill, 2000).

James Cherry

Vice-Chair of the BoG, James Cherry is the President and CEO of Aéroports de Montréal, with a long history in the aerospace and defence industries. Of note, Cherry has served as President and General Manager of Bombardier's Amphibious Aircraft Division and on the Mitec Telecom Inc. Board of Directors (Mitec, 2004). Bombardier boasts a state-of-the-art Military Training Program providing "innovative and cost-effective integrated training solutions, producing world-class military aircrew" (Bombardier, 2011). In 2009 Mitec, a "leading designer and manufacturer of mobile wireless, fixed wireless, broadcast and satellite components", was awarded $400,000 from a U.S. military customer for a specially-designed high-powered pulse amplifier (Mitec, 2009). In November of 2010 they announced a $1 million contract with the Canadian military for radar system components (Mitec, 2010).

Brian Edwards

Brian Edwards sits as Vice-Chair on the BoG as well as sitting on the Camoplast Board of Directors. Camoplast manufactures rubber tracks for military applications, "providing the most mature technology available for mission survivability in hostile environments" (Camoplast, 2011). Among Camoplast's "successfully fielded products" is the TALON™-SWORDS robot for Special Operations Command/U.S. Army, a "powerful, lightweight, versatile robot designed for missions ranging from reconnaissance to weapons delivery" (Global Security, 2005, ¶5).

L. Jaques Ménard

L. Jaques is the Chancellor of the BoG, the Chairman of BMO Nesbitt Burns and the President of BMO Financial Group, Québec. The Bank of Montréal (BMO) is the official bank of the Canadian military community. The Canadian Defence Community Program is "designed for the unique needs of the military community," and is available to "Regular CF and Reserve Force members, retired and former CF members, DND employees, NPF employees, and family members" (Canadian Forces Personnel Support Agency [CFPSA], 2009). BMO has also launched a "Support Our Troops" MasterCard, making it easy for everyday Canadians to fund the Canadian military above and beyond federal tax dollars (CFPSA, 2009).

Ivan Velan
Ivan Velan is the Executive Vice-President, North American Sales, Quality Assurance and MIS of Velan Incorporated. Velan Inc. is "a leader in valves for the marine industry, having supplied valves to more than 950 U.S. Navy and NATO ships, and all U.S. Navy nuclear submarine and aircraft carriers—most recently to the latest U.S. nuclear aircraft carrier, the USS Gerald R. Ford" (Velan, 2011).

Ethical Implications and Considerations

As Cleve Higgins has noted in his Honours thesis on the development of explosives technology at McGill, "when military agencies support university research, there is necessarily a potential application of the research that has military significance" (Higgins, 2009, p. 5). The research undertaken by Canadian universities that has been outlined above has real-life implications at home and on the battlefield. Military technology developed in part by Canadian professors and students increases the killing capacity of Canadian, U.S. and Israeli militaries, contributing to the military operations, wars and human rights violations in which these countries are involved. Even when not directly funded by the U.S. military, technology developed by Canadian universities can contribute to U.S. weapons development, and thereby the expansion of the American empire. For example, McGill SWPG papers have been used as background references for a U.S. Air Force technology contract solicitation for "Methods to Direct and Focus Blast" for SFAE bombs (USAF, 2005), indicating that "the research happening at McGill was contributing to U.S. weapons development, even though it was decades since the research had been funded by the U.S. military" (Higgins, 2009, p. 32). Moreover, "most major Canadian universities, including the University of Toronto, have received funding from various American military agencies (e.g. U.S. Navy, U.S. Air Force, U.S. Strategic Defense Initiative)," as revealed by the Canadian Military Industry Database Report (see Kim, 2002, ¶ 3).

UAV technology represents an incredible increase in the killing capacity of the U.S. military while keeping U.S. military personnel safe from harm. In addition to the deadly consequences of this technology on the battlefield, UAVs have been criticized for making it psychologically easier for soldiers to kill humans. Operation Objection asserts that, "autonomous weapons decouple the killer from the killed; compared to a conventional combat soldier or police officer, the deployer of a robot weapon faces reduced personal risk and may feel reduced responsibility for deaths that result" (Operation Objection, 2008, p. 12).

Academic Ethical Research Policy

The Canadian Association of University Teachers emphasizes the role of education towards the common good of society, stating that "education's most basic purpose is to enhance life and the dignity of the human person, an objective that is difficult to achieve in the absence of fundamental human rights" (Canadian Association of University Teachers [CAUT], 2002, ¶ 2). The association even has a policy regarding private donations towards research which asserts that "institutional governing bodies should review the activities and the record of prospective private funders and reject a funding source if it has violated human rights or behaved unethically" (CAUT, 2006, ¶ 5). In practice, however, university ethical research policies do not extend past the immediate ramifications of research. There are currently no policies regarding the ethical implications of military research outside of the university sphere at any university in Canada. According to a letter written by McGill Vice-President Research and International Relations Denis Thérien to his counterpart Ted Hewitt at University of Western Ontario following a discussion with 11 other Vice-Presidents of Research from Canadian universities on military-funded research in 2007, "no institution currently undertakes, or is contemplating undertaking any formal assessment of military projects on ethical or other grounds not already stipulated by existing guidelines. Nor would we endorse the creation of any national body to establish guidelines for such a Process" (quoted in Law, 2010a, ¶ 9). The only university that once had a policy of this nature was McGill University. The policy was established in 1988 following student campaigns against FAE research but was revoked in 2010, three years after Thérien's letter to Hewitt (Higgins, 2009; Law, 2010b).

The question of the institutionalization of an ethical review of military research was brought up at Western University following a concerted campaign against a research contract between Western's Department of Engineering and military industrial corporation General Dynamics. The contract was towards the development of the Stryker Light Armoured Vehicle, a multipurpose military vehicle employed by the U.S. Army in Iraq (Dyer-Witheford, 2007). The Western University Research Board, the Ontario Council on University Research and the Vice Presidents of Research for the G-13 research universities in Canada unanimously rejected the idea of an ethical review of military research, stating that a review process would be both impractical and against the principle of academic freedom (Operation Objection, 2008).

Demilitarization Campaigns

While ignored by Canadian university administrations and faculties, the ethical implications of military research are a focus of organizations opposing the militarization of Canadian universities. These organizations include Anthropologists for Justice and Peace, Science for Peace, Anti Recruitment, Operation Objection, People against the Militarization of the Ontario Institute for the Study of Education, Demilitarize McGill, Solidarity for Palestinian Human Rights and Tadamon. Since the 1960s there have been numerous campaigns across the country organized by these, and other, groups against specific military research undertaken by Canadian professors and students. These groups are quick to point out the ramifications of military research and technology outside of the university sphere and the need for transparent research and institutionalized processes for considering ethical implications. Demilitarize McGill asserts that "for decades McGill has been producing knowledge that enhances the killing capacity of the U.S. military. The details of this, and any other military research, must be made transparent so that there can be a formal, public discussion at the university of its ethical implications" (Demilitarize McGill, 2011).

In a presentation to the University of Western Ontario's University Research Board on May 8, 2007, Nick Dyer-Witheford of Science for Peace spoke to the ethical implications of military research as well as the obligation of the university to faculty, staff and students implicated in such research:

> "A university researching for the military, or for a military contractor, must reckon on the possibility that this research will help kill someone. Moreover...this killing will not necessarily have been approved by our national government. In the case of General Dynamics, it is killing in a war that Canada did not join. In the case of other United States defense contacts, it could be killing in Iran, Venezuela or Columbia, and, given the internationalized aspect of the arms trade, it is not impossible that in future UWO [University of Western Ontario] researchers could be working on weapons contracts for European, Chinese or Russian companies whose weapons would be used in Sudan, Chechnya or Tibet....The university has a responsibility to ensure that this process [of research] does not, even if indirectly, implicate students, staff or faculty in violations of international law or offences against human rights.[...] The graduate students today working on Strykers, or any other military related project, may in the future have to come to terms with the fact that in their youth they worked on supplying armaments for one of the first great humanitarian catastrophes of the twenty-first century. We have a responsibility to help prevent such nightmares, for the sake of both the perpetrators and the victims." (Dyer-Witheford, 2007, ¶ 6-9)

Conclusion

The corporatization of Canadian universities through federal funding cuts, privatization of research and an emphasis on public-private partnerships has been extremely conducive to the processes of militarization. Canadian professors and students all over the country contribute to the creation of weapons of war, the perpetuation of military ideologies and the expansion of the American Empire while Canadian academic administrations and faculty boards remain removed from discussion of the ethical implications of military research. The focus on short-term implications of research, without thought to the application of military research and technology outside of the classroom is a grave failure on the part of these administrative bodies and on the part of professors and students involved in such research. This failure, combined with the lack of transparency of military research makes it difficult, if not impossible to have an open dialogue on ethical implications within the university community. There is a pressing need for the institutionalization of processes of ethical review that take into consideration the full gambit of the implications of military research. Sadly, the cases of McGill University and the University of Western Ontario do not provide much hope in this direction. There is also a pressing need for a greater conversation about the role of the university: is the university to be a bastion for critical thinking and civil responsibility, or a manufacturer of state and industrial needs? Until the time that these issues are addressed by the community-at-large the Canadian military-academic-industrial complex will continue to thrive; Canadian universities have indeed become instruments of state and military interests. In his farewell address, President D. Eisenhower stated that, "Disarmament, with mutual honour and confidence, is a continuing imperative. Together we must learn how to compose differences, not with arms, but with intellect and decent purpose" (Eisenhower, 1961, p. 1039). To the detriment of the integrity of Canadian academic institutions, it seems as though this imperative has been ignored.

Notes

* A shorter and earlier version of this chapter was published as "The Militarization of Canada's Universities: From simulation programs to unmanned drones, Canada's schools have joined the fight," in *The Mark* on 01 June 2011—see: http://www.themarknews.com/articles/5387-the-militarization-of-canada-s-universities?page=1; it was also published on the WikiLeaks news website, *WL Central*, on 03 June 2011—see: http://wlcentral.org/node/1842; and, it was also published under that title on the site of *Anthropologists for Justice and Peace* on 15

June 2011—see: http://anthrojustpeace.blogspot.com/2011/06/militarization-of-canadas-universities.html.

References

Aviya. (2008). Aviya Technologies Awarded DO-254 Certification Contract. Aviya Tech News, May 15.
  http://www.aviyatech.com/news.php?id=4#Article

Bell Helicopter Textron. (2011). Military Aircraft: Your Mission: Preserve Freedom.
  http://www.bellhelicopter.textron.com/en_US/Military/Military.html

Bernans, D. (2001). Con U Inc.: Privatization, Marketization and Globalization at Concordia University (and beyond). Montréal: Concordia University Student Union.

Bombardier. (2011). Military Training.
  http://www.bombardier.com/en/aerospace/services-and-solutions/customer-support-and-services/training-services/military-training?docID=0901260d8000dc09

Bowen, D.G. & MacKenzie, S.C. (2003). Autonomous Collaborative Unmanned Vehicles: Technological Drivers and Constraints. Defence Research Development Canada Contract Report: DRDC CR-2003-003, September.
  http://pubs.drdc.gc.ca/PDFS/unc28/p519508.pdf

Camoplast. (2011). Camoplast—Track System Solutions.
  http://www.camoplast.com/en/tracksystem/replacementtracks_defense_all.php

Canadian Association of University Teachers (CAUT). (2002). CAUT Policies: Policy Statement on Social Justice. November.
  http://www.caut.ca/pages.asp?page=297&lang=1

————— . (2006). CAUT Policy Statement Guidelines for Acceptance of Private Donations. September.
  http://www.caut.ca/pages.asp?page=277&lang=1

Canadian Forces Personnel Support Agency (CFPSA). (2009). A program designed for the unique needs of the military community. September.
  http://www.cfpsa.com/en/corporate/Publications/MWTeamUpdates/September2009/bmo_e.asp

Centre d'Étude des Politiques Étrangères et de Securité (CEPES). (2011). Home Page.
  http://www.cepes.uqam.ca

Centre for Environmental Research in Minerals, Metals and Materials (CERM3). (2011). Dr. John A. Meech.
  http://www.cerm3.mining.ubc.ca/workshop/meech.htm

CITIC Pacific. (2011). Home Page.
  http://www.citicpacific.com/

Concordia University. (1999). Concordia University Awards for Research Grants and Contracts. November 24, obtained through Concordia Student Union access to information request.

————— . (2011a). Faculty of Engineering and Computer Science: Home Page for Dr.

Youmin Zhang.
    http://users.encs.concordia.ca/~ymzhang/index.html
———— . (2011b). History–Concordia University. Montreal, QC: Concordia
    University.
    http://www.concordia.ca/about/who-we-are/history/
Concordia University Board of Governors (BoG). (2011). Office of the Vice-President,
    External Relations and Secretary-General, Concordia University, Montréal,
    Québec, Canada.
    http://www.concordia.ca/vpirsg/board-and-senate/governors/
Concordia Institute of Aerospace Design and Innovation (CIADI). Projects for
    Summer and Winter 2011.
    http://ciadi.concordia.ca/projects.htm
Concordia's Thursday Report. (2001). Current Concordia University Mission
    Statement. Concordia's Thursday Report Online, November 8.
    http://ctr.concordia.ca/2001-02/Nov_8/27-Of_Note/index.shtml
Concordia Senate. (1999). Minutes of the Meeting of 17 September 1999. Document
    US-99-5.
Defence Research and Development Canada (DRDC). (2010). Partnering with
    Defence Research and Development   Canada.
    http://www.drdc-rddc.gc.ca/partner-partenariat/prog/nserc-crsng/index-eng.asp
Defense World. (2011). CAE wins C$140 million contract for C-130J weapon systems
    from Canada. DefenseWorld.net, January 21.
    http://www.defenseworld.net/go/defensenews.jsp?id=5379&h=CAE%20wins%2
    0C$140%20million%20contract%20C-
    130J%20weapon%20systems%20from%20Canada
Demilitarize McGill. (2011). Demilitarize McGill: For Transparent and Ethical
    Research.
    http://demilitarizemcgill.wordpress.com/
Dyer-Witheford, N. (2007). Military Related Research at the University of Western
    Ontario. Science for Peace Bulletin, June 20.
    http://www.scienceforpeace.ca/0707-military-related-research-at-uwo
Eisenhower, D.D. (1961). Dwight D. Eisenhower: 1960-61: Containing the Public Messages,
    Speeches, and Statements of the President, January 1, 1960, to January 20, 1961. (pp.
    1035-1040). Ann Arbor, Michigan: University of Michigan Library.
    http://quod.lib.umich.edu/p/ppotpus/4728424.1960.001?view=toc
EnCana. (2011). Fourth Quarter and year-end 2010 Results–February 10, 2011.
    http://www.encana.com/investors/financial/quarterly/
Forte, M.C. (2010). Introduction: The "New" Imperialism of Militarization,
    Humanitarianism, and Occupation. In Maximilian C. Forte (Ed.), The New
    Imperialism, Volume 1: Militarism, Humanism, and Occupation (pp. 1-29). Montréal,
    QC: Alert Press.
Fulbright, W.J. (1970). The War and Its Effects: The Military-Industrial-Academic
    Complex. In H.I. Schiller & J.D. Phillips (Eds.), Super-State: Readings in the
    Military-Industrial Complex (pp. 173-178). Urbana, IL: University of Illinois Press.
Giroux, H. (2007). The University in Chains: Confronting the Military-Industrial-Academic
    Complex. Boulder, CO: Paradigm Publishers.

————— . (2009). Making Democracy Matter: Academic Labor in Dark Times. *Counterpunch*, March 11.

http://www.counterpunch.org/giroux03112009.html

Global Security. (2005). Talon Small Mobile Robot. http://www.globalsecurity.org/military/systems/ground/talon.htm

Higgins, C. (2009). From Bibliographies to Battlefields: Military Research at McGill from 1967-2007. Sociology Honours Thesis, Supervised by Professor Donald Von Eschen.

Industry Canada. (2011). Canadian Aerospace and Defence Industry—Industrial Regional Benefits (IRB) Policy.

http://www.ic.gc.ca/eic/site/042.nsf/eng/home

Johnson, C. (2000). *Blowback: The Costs and Consequences of American Empire*. New York: Henry Holt and Company.

Kim, B., Johnson, B., Youssef, R., Vallerand, A.L., Herdman, C., Gamble, M., Lavoie, R., Kurts, D. & Gladstone, K. (2005). JSMARTS Initiative: Advanced Distributed Simulation Across the Government of Canada, Academia and Industry—Technical Description, July 2005. Technical Memorandum. Ottawa, ON: Defence Research and Development Canada.

http://handle.dtic.mil/100.2/ADA439997

Kim, H. (2002). Science for Peace: Military Research and Canadian Universities. *Science for Peace Bulletin*, September 20.

http://www.scienceforpeace.ca/0209-military-research-and-canadian-universities

Lacome, J.L. (1998). Pledge letter from Pratt and Whitney to Richard J. Renaud, Chair of the Concordia University and Friends Division of Annual Giving. Obtained through Concordia Student Union access to information request.

Law, S. (2010a). Speaking Out Against Military Research Since the 80's: CKUT Speakers Discuss Current Policy's Failure to Regulate Harmful Research Applications. *The McGill Daily*, March 15.

http://www.mcgilldaily.com/2010/03/speaking_out_against_military_research_since_the_80s/

————— . (2010b). McGill Will Not Regulate Military Research: After a Year-Long Battle in Senate, Harmful Research Disclosure Left Out. *The McGill Daily*, April 12.

http://www.mcgilldaily.com/2010/04/mcgill_will_not_regulate_military_research_/

Locke, J., & Whelan, M. (2011). RDC and Boeing investments help create new engineering facility. *Memorial University Today*, March 28. http://today.mun.ca/news.php?news_id=6269

Maskery, M., Krishnamurthy, V., & O'Regan, C. (2005). Game Theoretic Missile Deflection in Network Centric Warfare. Presented at the 2005 IEEE International Conference on Networking, Sensing and Control. Ottawa, ON: Defence Research and Development Canada.

http://pubs.drdc.gc.ca/PDFS/unc44/p524656.pdf

Medeiros, E.S., & Gill, B. (2000). *Chinese Arms Exports: Policy, Players and Process*. Carlisle, PA: Strategic Studies Institute, U.S. Army War College.

Meister, J.C. (1998). *Corporate Universities: Lessons in Building a World-Class Work Force*. New York: McGraw-Hill.

Mitec. (2004). Mitec Appoints David E. Scott and James C. Cherry to its Board of Directors. *Mitec Telecom Inc.*, June 3.

http://www.mitectelecom.com/press/file_en_121.pdf

———. (2009). Mitec Begins Delivery of New Military Amplifier Line. *Mitec Telecom Inc.*, October 26.

http://www.mitectelecom.com/press/mitec_begins_delivery_of_new_military_a mplifier_line_-_october_26,_2009_en.pdf

———. (2010). Mitec Awarded $1 Million Canadian Military Contract. *Mitec Telecom Inc.*, November 30.

http://www.mitectelecom.com/press/mitec_awarded_$1_million_canadian_navy _contract_-_november_30,_2010_english.pdf

National Defence and the Canadian Forces. (2010). Academic Programs—Security and Defence Forum.

http://www.forces.gc.ca/admpol/SDF-eng.html

Operation Objection. (2008). Military Research in Our Universities: An Analysis of Academic Research for the Canadian Army. Montreal, QC: Operation Objection.

http://www.antirecrutement.info/files/Military%20research%20in%20universiti es%20-%20Operation%20Objection.pdf

Public Works and Government Services Canada. (2011). Contract History.

http://csi.contractscanada.gc.ca/index-eng.cfm?af=ZnVzZWFjdGlvbj1pbmZvLnN0YXJ0JmlkPTE=&lang=eng

Security Defence Forum (SDF). (2008). *Security Defence Forum: Year in Review 2007–2008*. Ottawa, ON: National Defence and the Canadian Forces.

http://www.forces.gc.ca/admpol/annualreport2008.html

Tether, T. (2006). Statement by Dr. Tony Tether, Director Defense Advanced Research Projects Agency. Submitted to the Subcommittee on Terrorism, Unconventional Threats and Capabilities House Armed Services Committee United States House of Representatives on March 29, 2006. http://www.darpa.mil/WorkArea/DownloadAsset.aspx?id=1769

Thales. (2011). Supporting the Forces in their New Missions. *Thales Group*.

http://www.thalesgroup.com/Markets/Defence/Home/

Tudiver, N. (1999). *Universities for Sale: Resisting Corporate Control over Canadian Higher Education*. Toronto: James Lorimer and Company.

Turse, N. (2004). The Military-Academic Complex. Z Net, April 27.

http://www.zcommunications.org/the-military-academic-complex-by-nick-turse.pdf

UBC Thunderbird Robotics. (2011). Home Page.

http://www.ubcthunderbird.com/

United States Air Force (USAF). (2005). Methods to Direct and Focus Blast. AF05-153, Munitions Directorate, Eglin AFB. Eglin: Munitions Directorate.

http://www.dodsbir.net/SITIS/archives_display_topic.asp?%20Bookmark=2002 6

Vallerand, A.L., Kim, B., Youssef, R., Hubbard, P., Skinner, D., Murray, B., Poursina, S., Herdman, C.M., Gamble, M., Hagen, L., Bleichman, D., Kurts, D., Kruk, R., Lavoie, R., & Gladstone, K.G. (2005). Synthetic Environments at the Enterprise Level: Overview of a Government of Canada (GoC), Academia and Industry

Distributed Synthetic Environment Initiative, April 2005. Ottawa, ON: Defence R&D Canada.

http://www.dtic.mil/cgi-bin/GetTRDoc?AD=ADA439998&Location=U2&doc=GetTRDoc.pdf

Vancouver Institute for Visual Analytics (VIVA). (2011). Boeing Visual Analytics Research Grant. Simon Fraser University: Vancouver Institute for Visual Analytics.

http://viva.sfu.ca/va-research/canadian-research/boeing-va-grant/

Véhicule Aérien Miniature de l'Université de Sherbrooke (VAMUdeS). (2007). Home Page.

http://mecano.gme.usherb.ca/~vamudes/frame_central_anglais/home.html

Velan. (2011). Industries Served: Marine.

http://www.velan.com/velan/en/industries_served?id=135

Vincelli, M. (2000). Concordia Launches Aerospace Institute. *Concordia's Thursday Report*, November 23.

http://ctr.concordia.ca/2000-01/Nov_23/01-CIADI/index.shtml

CHAPTER TWO

A WAR OF WORDS:
U.S.-IRAN RELATIONS IN THE NUCLEAR DEBATE

Sabrina Guerrieri

"We have come together and founded cities and made laws and invented arts; and, generally speaking, there is no institution devised by man which the power of speech has not helped us to establish".— Isocrates, *Antidosis*

The United States of America and the Islamic Republic of Iran as two geopolitical powers—albeit one global and the other regional—find themselves in a situation of mutual hostility, or in military terms, an asymmetrical conflict. Indubitably, the word "confrontational" would be most apt in summarizing U.S.-Iran relations, which far from being a new phenomenon, is the result of a relatively extended history of "perceived betrayals, open challenges and reciprocated distrust" (Cook & Roshandel, 2009, p. 1). Among the numerous layers that construct the foundation of such an antagonistic liaison, this chapter will have as its focus the particularly significant battle over Iran's nuclear program. Mildly put, this topic is overwhelmingly broad and has been subject to the analysis of numerous scholars who have explored divergent channels. The pages that follow will thus not aim attention at the factors that make Iran different from other states in the realm of proliferation or military theory, such as the reasons why it may even need to obtain a bomb so desperately. Neither will the following pay tribute to the history of Iran's specific nuclear proliferation program which began in 1975 under the Shah, Mohammad Reza Pahlavi, an ally of the U.S.-UK sponsored coup against Prime Minister Mohammad Mossadeq (Kahn, 2010, p. 47). Despite the vital importance of these domains in understanding the full conflict, the focus of this chapter will reside in exploring a route often ignored by analysts: the national rhetoric used by officials on both sides of the equation that contribute to the persisting negative view of the "other" and in turn, how it translates into public opinion. Rhetoric, for the purpose of this essay, will be understood as the persuasive use of language that has the capability of influencing civic life. In line with the claims of the ancients, modern thinkers such as the literary critic and American Law professor, James Boyd White (1984), argue that rhetoric is capable not only of addressing issues of political interest but that it can influence culture as a whole. White states that words and language "produce, acquire and hold and lose their meanings, with the methods by which culture is maintained, criticized, and transformed" (1984, p. 279).

This chapter will predominantly but not exclusively centre on post 2005 events for two main reasons. First, it is the year Iran's controversial president Mahmoud Ahmadinejad took office and has since then, been strongly associated with the country's perceived desire to acquire nuclear weapons (Kahn, 2010, p. 14). Therese Delpech, the director of strategic studies at the French Atomic Energy Commission, explains:

> "Mahmoud Ahmadinejad's arrival in power, his inflammatory speeches, his determination to plow through at all costs and the feebleness of the international response to these developments all reinforce the skepticism and even the fear of some future catastrophe". (2006, p. 37)

Second, 2005 was the year in which Iran was declared to be "unequivocally in violation of its obligations in a Board of Governors resolution which cites article 12 of the International Atomic Energy Agency (IAEA) status, making it compulsory for a report to be submitted to the Security Council" (Delpech, 2006, p. 91). Prior to 2005, the international community and the IAEA seemed to be hesitant in recognizing Iran to be in violation of agreements (Kahn, 2010, p. 14). Accordingly, the following pages will first present a brief overview of the debate, in turn followed by how each nation perceives the other and consequently how they chose to deal with those perceptions. In conclusion, the chapter will bring forth the potential of alleviating the tensions within U.S.-Iran relations with the aim of ultimately forming a more constructive relationship.

Two Sides of the Debate

Iran's nuclear debate, when extirpated from its ornaments, rests on the concept of sovereignty, that is, within the modern notion of supremacy of authority exercised by a sovereign state. Just as Iran argues it is its sovereign right to produce nuclear energy for peaceful purposes, the U.S. argues it is its sovereign right to dissuade countries from attaining nuclear technologies (Cook & Roshandel, 2009, p. 117). Such nuclear dissuasion however, can be criticized as being biased—as being a selective process applied towards countries perceived as a threat to the international community. Iran, as a prominent target in this list, has consistently emphasized its nuclear program as a peaceful one, with a purpose of satisfying civilian energy needs and guiding the country towards economic growth (Cook & Roshandel, 2009, p. 118). Although Iran does hold roughly 20% of the world's total oil reserves with an estimated 157.6 billion barrels of proven oil reserves, Iran's crude oil production capacity was estimated to be a mere 3.9 million barrels per day (Energy Information Administration [EIA], 2010). Iran is the fourth largest crude oil exporter in the world (primarily to Asia and OECD Europe countries). However, it is unable to keep pace with domestic demand for diesel and gasoline (EIA, 2010). Iran's limited refinery capacity for the production of such light fuels makes it dependent on the importation of refined oil, hence the need for the diversification of energy sources. It must be emphasized that it is a sovereign right for all countries to strive for energy independence to which this pursuit affords, and that international organizations should facilitate such progress rather than impose barriers. By being a member of the Nuclear Proliferation Treaty (NPT), which it signed in 1968 and ratified in 1970 (Kahn, 2010, p. 12), Iran does indeed have the right to nuclear technologies, albeit for peaceful purposes only. Article IV of the NPT (2000) states:

1. Nothing in this Treaty shall be interpreted as affecting the inalienable right of all the Parties to the Treaty to develop research, producing and use of nuclear energy for peaceful purposes without discrimination and conformity with Articles I and II of this Treaty
2. All the Parties to the Treaty undertake to facilitate, and have the right to participate in, the fullest possible exchange of equipment, materials and scientific and technological information for the peaceful uses of nuclear energy.

In light of these NPT rights, Iran accuses the U.S. of attempting to prohibit it from having peaceful nuclear capabilities (Cook & Roshandel, 2009, p. 118). Seeing that Iran has a peculiar history of anti-Western and anti-American sentiments beginning before the Islamic Revolution of 1979, U.S. actions towards the nuclear issue are interpreted as yet another way that the West denies other countries their rights in the international system. In fact Article 152 under Chapter X of the Islamic Republic of Iran's Constitution (1979) illustrates a staunch rejection of any form of imperialism or foreign interference in domestic affairs, particularly from the world's biggest superpower:

> "The foreign policy of the Islamic Republic of Iran is based upon the rejection of all forms of domination, both the exertion of it and submission to it, the preservation of the independence of the country in all respects and its territorial integrity, the defence of the rights of all Muslims, nonalignment with respect to the hegemonist superpowers, and the maintenance of mutually peaceful relations with all non-belligerent States".

The NPT also allows the right of any sovereign state to withdraw their membership (as did North Korea in 2003), something Iran has not done as a gesture of their proclaimed peaceful intentions (Cook & Roshandel, 2009, p. 123). On 10 August 2005 the Iranian nuclear negotiator at the time, Sirus Naseri, read a statement to the IAEA on behalf of the Iranian Supreme Leader, which asserted the religious denunciation of nuclear weapons based on the beliefs of Shia Islam:

> "The Leader of the Islamic republic of Iran, Ayatollah Ali Khamenei has issued a fatwa [a religious edict] that the production, stockpiling and use of nuclear weapons are forbidden under Islam and that the Islamic Republic of Iran shall never acquire these weapons". (Mehr News Agency, 2005, ¶ 9)

Khamenei also asserted in 2008 that although "no wise nation" would pursue nuclear weapons, Iran would continue to develop a nuclear program for peaceful purposes (Cordesman & Seitz, 2009, p. 9). The debate thus shifts from a theoretical understanding of sovereignty to a much more pragmatic disagreement: is Iran's nuclear program an attempt to increase its energy production as it claims or is it seeking nuclear weapons?

The U.S. position stands firmly on the latter and not without reason. While the U.S. claims to acknowledge Iran's sovereign right to peaceful nuclear power, it argues that the size and configuration of this facility is at odds with that of a peaceful program (Obama, Sarkozy, & Brown, 2009, ¶ 4). Moreover, IAEA inspections have presented substantial evidence that Iran has been engaged in undeclared nuclear activities for over two decades (Cook & Roshandel, 2009, p. 130), which has instigated the suspicion of their nuclear intentions. The problem according to NPT agreements is that Iran has advanced its nuclear program covertly and only placed its facilities under IAEA safeguards once they were detected by the international community (Cook & Roshandel, 2009, p. 130). Additionally, IAEA evidence revealing Iran's insistence on the acquisition of heavy water reactors and highly enriched uranium production (see IAEA, 2009) can only be explained, argues the U.S., by a nuclear weapons aspiration (Cook & Roshandel, 2009, p. 130). The IAEA has many a times asserted that Iran continues to violate the directives and decisions of the UN Security Council. For instance the IAEA Board of Governors stated:

Figure 2.1

Grand Ayatollah Ali Khamenei, the current Supreme Leader of Iran. (Image provided by Wikimedia Commons under a Creative Commons License.)

"Contrary to the decisions of the Security Council, Iran has not suspended its enrichment related activities, having continued the operation of PEEP and FEP and the installation of both new cascades and of new generation centrifuge for test purposes. Iran has also continued with the construction of the IR-40 reactor 9". (IAEA, 2007, p. 8)

This has consequently affected the trust of the international community, or as Obama worded it at the September 2009 G-20 summit: "It is time for Iran to act immediately to restore the confidence of the international community by fulfilling its international obligations" (Obama, Sarkozy, & Brown, 2009, ¶ 5). The U.S. has worked through the UN Security Council to try and pressure Iran to turn back from this plan and make its nuclear program more transparent (Cook & Roshandel, 2009, p. 118). Before the 65th Session of the United Nations General Assembly on September 23rd 2010, Ahmadinejad defended Iran's case on the ground of corruption within the Security Council itself, whereby the permanent members abuse their hegemonic power:

> "Nevertheless, note what some of the permanent members of the Security Council and nuclear bomb holders have done: they have equated nuclear energy with the nuclear bomb, and have distanced this energy from the reach of most of nations by establishing monopolies and pressuring the IAEA. While at the same time, they have continued to maintain, expand and upgrade their own nuclear arsenals". (Ahmadinejad, 2010, p. 5)

As a perfectly reasonable statement, one could assume that Ahmadinejad was addressing the U.S. and its ally, the state of Israel, which is understood to have the Middle East's only nuclear arsenal, a fact that has never officially been confirmed or denied (Dahl & Westall, 2010, ¶ 5). On a broader level, Ahmadinejad's statement was a direct challenge to the role of major world institutions such as the UN and the IAEA which, as institutions of the new imperialism, serve to secure the economic and ideological dominance of U.S. imperial power. In just a brief overview of the Iranian nuclear debate, the tensions and frictions found within U.S.-Iran relations quickly become apparent. The following paragraphs will review some of the main narratives projected by both the U.S. and Iran, which will clearly illustrate to what extent such hostility has become naturalized in the cultural and political narratives of one another.

The U.S. View of the Other

In one of his most vital chapters, directly titled "U.S. Policy and Approaches Regarding Iran," Robert E. Hunter in a RAND publication, states: "Iran's attitude and role may be the most problematic question of all in regards to the potential for creating a viable Persian Gulf regional security structure" (2010, p. 41). U.S. strategy building towards Iran, Hunter states, will largely depend on whether it continues to be uncooperative or relatively cooperative with the majority parts of the outside world (2010, p. 36). It would be safe to assume that the U.S., for quite some time, has been maneuvering its foreign policy on the assumption that Iran will remain largely uncooperative. For instance, according to the IAEA, from 2000 to 2005, Iran did reinforce its efforts to build nuclear weapons under the cover of a civilian nuclear program, which gravely influenced U.S. foreign policies towards Iran (Kahn, 2010, p. 90). However as Ali Ansari, one of the world's leading experts on Iran and its history states (2006), one cannot disassociate U.S. actions from Iranian actions or vice versa: they must be understood as interdependent. Ansari elaborates: "The growing confrontation over Iran's nuclear program could not be understood outside the general political malaise that characterized Iran-U.S. relations" (2006, p. 201). The official rhetoric of the U.S. has been to cast Iran as the "bad guy", i.e. a "threat" to the international community should they continue what the U.S. is convinced is a nuclear weapons drive. In his famous 2002 State of the Union "Axis of Evil" speech, President George W. Bush asserted that Iran, Iraq and North Korea were "arming to threaten the peace of the world":

> "Iran aggressively pursues weapons [of mass destruction] and exports terror, while an unelected few repress the Iranian people's hopes from freedom....States like these, and their terrorist allies, constitute an axis of evil, arming to threaten the peace of the world. By seeking weapons of mass destruction, these regimes pose a grave and growing danger. They could provide these arms to terrorists, giving them the means to match their hatred". (Bush, 2002, ¶ 19 & 21)

The Bush administration along with its neo-conservative policy-making circle viewed military coercion as the most effective foreign-policy strategy in combating such rogue states and their proliferation activities (Kahn, 2010, p. 92). In 2003, the U.S. sent clear signals to countries like Iran that it would wage wars on countries similar to Iraq (anti-Western states) attempting to violate global security norms (Kahn, 2010, p. 92). Such rhetoric was thus vigorously pushed through American mainstream media. A golden example is the Fox News special, bluntly titled "Iran: The Nuclear Threat," hosted by Chris Wallace, which was meant to persuade American viewers on the authenticity of the Bush administration's "War on Terror". Despite the numerous traces of war propaganda, the so-called "experts" reassured America that the allegations against Iran were indeed objective truths. Yet as Henry Kissinger has pointed out, in this "age of the expert" the constituency of the expert consists of, "those who have a vested interest in commonly held opinions; elaborating and defining its consensus at a high level has, after all, made him an expert" (1969, p. 28). Even with the change of the American administration in 2009, little "change" has come into effect. Along the same lines as Bush only with more sophisticated adjectives, President Obama has cast Iran not only as a "threat" but as an "unusual and extraordinary threat" due to its persistent unwillingness to meet its obligations under U.N. Security Council resolutions and IAEA requirements. Obama has stated:

> "Because the actions and policies of the government of Iran continue to pose an unusual and extraordinary threat to the national security, foreign policy, and economy of the United States, the national emergency declared on March 15, 1995, must continue in effect beyond March 15, 2011". (Obama, 2011, ¶ 2)

As Noam Chomsky and Edward S. Herman emphasize in *Manufacturing Consent*, the White House, Pentagon, and State Department in Washington, DC, are central figures in dictating the exportation of "objective" news (1988, p. 19). As a result, one might question how such a narrative of Iran as the "bad guy" manufactures American public consent?

Studies show that most Americans view Iran as a "serious threat" to the U.S. For instance the 2007 Pew Global Attitudes Project found an overwhelming majority of Americans (86%) believed that Iranian acquisition of nuclear weapons would be a "serious threat" to the U.S. (Gilboa, 2010, p. 3). In 2008, the Public Opinion Strategies poll found a similar majority of 87% expressing the same opinion and more recently, the 2010 Pew Global Attitude Project illustrated that 94% of Americans opposed Iran acquiring nuclear weapons (Gilboa, 2010, p. 3). And in the McLaughlin Group poll of October 2010, 86% of the respondents agreed that "Iran will use its nuclear weapons to threaten the supply of Middle Eastern oil to the U.S. and its allies"; only 10% disagreed (Gilboa, 2010, p. 4).

Iran's View of the Other

Ayatollah Khamenei's analysis of U.S.-Iranian relations is one that some find "strikingly similar to hardliners in Washington, who believe the two countries represent diametrically opposed ideologies destined for an inevitable confrontation" (Sadjapour, 2009, p. 15). Khamenei's address to Iranian government officials in May 2003, just after U.S. forces captured Baghdad, speaks volumes of a vision of intractable conflict:

"It is natural that our Islamic system should be viewed as an enemy and an intolerable rival by such an oppressive power as the United States, which is trying to establish a global dictatorship and further its own interests by dominating other nations and trampling on their rights". (Quoted in Sadjapour, 2009, p. 15)

Within the Iranian nation itself, the official narrative of the nuclear program has been molded into a tool of national pride and identity; exploiting concepts such as Iran's "rights" and the unjust burden of "double-standards" and "discriminations" imposed by the U.S. and its Western allies. As early as November 1991, Iran's Deputy President Ayatollah Mohajerani, stated that if the Zionist regime (i.e. Israel) has the right to possess nuclear weapons, then all Muslim countries should have that same right, thus implying that a double-standard ruled among the permanent members of the UN Security Council (Kahn, 2010, p. 23). Since then, voicing such a narrative has only intensified. In an exclusive interview with Fars news agency hosted by the Azeri Leader TV channel, Ahmadinejad has stated:

"Iran will not enter into talks with anybody on its indisputable nuclear right....No negotiation will yield positive results unless Western powers abandon their obsolete colonial mentalities. Baseless accusations against Iran are all leveled by those countries which possess giant nuclear stockpiles and at the same time keep arming their allies with such unconventional weaponry". (Press TV, 2010, 0:19)

This excerpt encompasses all three motifs, which together form a heightened rhetoric of victimization: 1) Iran's indisputable nuclear right; 2) the imposition of double-standards by the UN Security Council; and, 3) resentments of discrimination by the West. Likewise, Supreme leader Ali Khamenei's public narrative has tended to focus on the importance of science and technology in overcoming, what he refers to as Iran's "scientific retardation" which he claims, is the result of crippling restrictions by Western powers. In an address to High School students on 14 March 2005 Khamenei stated:

"It is hard for the global arrogance [of the U.S.] to accept that the talented Iranian nation has been able to take great strides in the field of science and technology, especially in the field of nuclear technology. They want Iran's energy to be always dependent on oil, since oil is vulnerable to the policies of world powers. They aim to control other nations with invisible ropes". (Quoted in Sadjadpour, 2009, p. 23)

Ironically such rhetoric coincides with the views of the Shah, under which the program initially began, who believed that a nuclear-development program would allow Iran to demonstrate that it was a modern, technologically competent country (Hunter, 2010, p. 34).

The 31st anniversary of the Islamic Republic of Iran was commemorated on 11 February 2010. On that day President Ahmadinejad addressed hundreds of thousands of government supporters in Tehran's Central Azadi Square, where he reiterated that Iran was "now a nuclear state" (Erdbrink & Kessler, 2010, ¶ 1). This was hardly a cause for celebration as similar statements have been proclaimed numerous times since his first announcement in April 2006 where the President first announced that Iran had succeeded in enriching uranium (Erdbrink & Kessler, 2010, ¶ 7). Ahmadinejad's chosen idiom made use of the plural "we," which cast himself as the spokesman for all citizens of the nation and reduced its heterogeneous population to a likeminded body of thought:

Figure 2.2

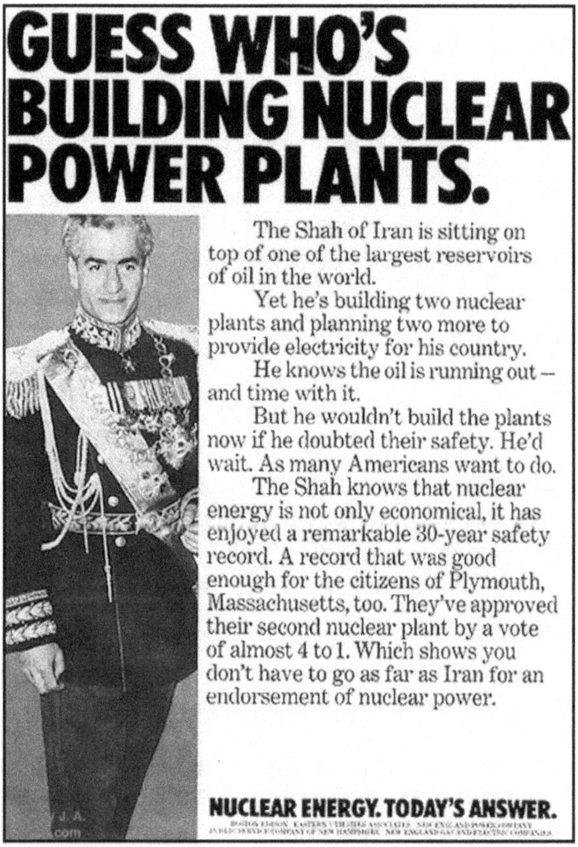

GUESS WHO'S BUILDING NUCLEAR POWER PLANTS.

The Shah of Iran is sitting on top of one of the largest reservoirs of oil in the world.

Yet he's building two nuclear plants and planning two more to provide electricity for his country.

He knows the oil is running out — and time with it.

But he wouldn't build the plants now if he doubted their safety. He'd wait. As many Americans want to do.

The Shah knows that nuclear energy is not only economical, it has enjoyed a remarkable 30-year safety record. A record that was good enough for the citizens of Plymouth, Massachusetts, too. They've approved their second nuclear plant by a vote of almost 4 to 1. Which shows you don't have to go as far as Iran for an endorsement of nuclear power.

NUCLEAR ENERGY. TODAY'S ANSWER.

A 1970s poster paid for by several U.S. nuclear energy firms, highlighting the Shah of Iran's nuclear program. (Image provided via Flickr under a Creative Commons License.)

"When we say that we don't build nuclear bombs, it means that we won't do that because we don't believe in having them....The Iranian nation is brave enough that if one day we wanted to create an atomic bomb, we would announce it publicly and would create it. We are not afraid of you". (Quoted in Erdbrink & Kessler, 2010, ¶ 5)

However, this "we" in regards to Iran's nuclear issue does not equate to the "national consensus" that it claims to reflect, for it excludes the divergent views (both within the political elite and the public) on how Iran's nuclear program should take course. The Supreme National Security Council, which embodies the President (whom actually wields little authority in matters of defense), the Defense and Foreign Ministers, the Commander of the Army of the Guardians of the Islamic Revolution (IRGC) and several appointees or "representatives" of the Supreme Guide (Chubin, 2009, p. 52), does not consist of a homogenous group and can be divided into three factions; the conservatives, the pragmatists and the reformers. Claiming "national consensus" gives negotiators power to act effectively and gets skeptical politicians on board (Chubin, 2009, p. 54). Sharam Chubin claims that Iranian public opinion is manipulated in order to support the program and prevent opposition:

"The nature of decision-making on a national security issues in Iran has tended to reflect the preference of the leadership to avoid public debate and to encourage citizen approval once a decision has been reached". (2009, p. 52)

What may appear to be a weak level of public participation is actually a leap forward; public opinion was never the slightest of an influencing factor prior to 2002, when the full program became exposed to the international community (Chubin, 2009, p. 53). Nonetheless, public opinion remains only a background factor in nuclear decision-making in the sense that certain public dissatisfactions cannot be ignored (Chubin, 2009, p. 53).

It is difficult to gauge how exactly the official narrative of Iran's nuclear program and views of the U.S. translate into Iranian public opinion; most scholars conclude that people are supportive of Iran's right to nuclear energy but agreement blurs on the cost they are willing to pay in terms of economic or political confrontations with the international community. Chubin states:

"While most Iranians agree on Iran's right to seek modern technologies unrestricted, consensus clearly fades over the price Iranians are willing to pay for the continuation of the program in terms of sanctions, loss of confidence in investment, capital flight and estrangement from the international community". (Chubin, 2009, p. 54)

A poll conducted in September 2010 by the University of Maryland's Program on International Policy Attitudes (PIPA) is one of few surveys which exemplify that the Iranian public is supportive of the nuclear program. When offered three alternatives in regard to Iran's nuclear program, (1) to develop both atomic bombs and nuclear power, (2) to only develop nuclear power, or (3) to have no nuclear programs, a modest majority of the general public (55%) wanted to develop nuclear power, only 38% wanted inclusion of weapons and 3% wanted no program at all (World Public Opinion [WPO], 2010, p. 20). Although one can clearly acknowledge a strong support of the nuclear program, as shall be later revealed, a portion of the population would favor the government precluding the development of nuclear weapons in exchange for the lifting of international sanctions against Iran (WPO, 2010, p. 20). Behind the existing consensus, some argue that there remains no informed debate about the energy rationale whereby alternative policies and options could be brought to the table (Chubin, 2009, p. 54).

Dealing with Iran: U.S. Foreign Policy

Although there has been a general consensus in the West on the threat Iran poses should it continue its nuclear program, there has been less agreement within the U.S. government on what should constitute its chosen foreign policy. Thus far, the U.S. and the UN have opted to dissuade Iran's enrichment activities through the imposition of numerous sanctions. The alleged benefit of sanctions is that they allow the U.S. government to express its discontent without having to commit U.S. military troops (Cook & Roshandel, 2009, p. 141). The U.S. defense of the application of sanctions against Iran is summarized below:

> "Iran's attempt to create a nuclear weapons capability, its sponsorship of terrorist groups in the Middle East and its atrocious human rights record, makes it one of the most urgent concerns for American foreign policy. Sanctions are an expression of American resolve to isolate the Government of Iran, increase diplomatic pressure and convince Iran to reverse its nuclear ambitions". (Cook & Roshandel, 2009, p. 141)

Proponents for the sanction strategies, argue it as a more "humane" option that would spare the West a war with Iran. Robert Gates, the U.S. Secretary of Defense until recently, believes that military action against Iran would only present a short-term solution (not to mention an array of blowback effects) and the only long-term solution is in the persistence of sanctions whereby Iranians themselves will have a change of hearts (Stewart, 2010, ¶ 10). This would occur due to mounting friction between ordinary Iranians and their government which is to blame for the isolation and its consequences. More recently, addressing an audience of West Point cadets, Gates stated: "In my opinion, any future defense secretary who advises the president to again send a big American land army into Asia or into the Middle East or Africa should 'have his head examined', as General MacArthur so delicately put it" (Shanker, 2011, ¶ 2).

Due to Iran's opaque government, there is much debate on whether sanctions placed on Iran are even achieving their intended goal or whether they are only fueling Iran's self-sufficiency. Author David Horovitz, presently editor-in-chief of *The Jerusalem Post*, reported that the sanctions are indeed beginning to "hurt Iran" by affecting the powerful IRGC (2010, ¶ 14) whose role it is to carry out a number of functions related to internal security, external defense and regime survival. The IRCG likewise fields an army, a navy and an air force (Green et al., 2009, p. 12). Washington's narrative has likewise been projecting a firm stance on the continuation of sanctions. Secretary of State Hillary Rodham Clinton said economic sanctions are slowing down Iran's ability to acquire nuclear weapons, and she urged Iran's neighbors to maintain pressure on the country (Kaufman, 2011, ¶ 1). While addressing university students in Abu Dhabi, Clinton said that "the most recent analysis is that sanctions have been working" and that they "have made it much more difficult for Iran to pursue its nuclear ambitions" (Kaufman, 2011, ¶ 2).

Yet Ron Paul, a Republican congressman, has publicly stated on the House floor on 23 April 2010 his disapproval with the Comprehensive Iranian Sanction bill, which was passed overwhelmingly by 400-11. Paul equated sanctions to an act of war against Iran and its allies, namely Russia and China, and that this was not in the long-term interest of the U.S.:

> "Sanctions are very dangerous; sanctions are literally an act of war. You prevent certain goods and services going in a country, it's like a blockade. There's no advantage for us to do this, the sanction bill literally says that any country that trades or sends oil into Iran, we will no longer trade with. So if Russia sends in oil or gasoline, refined products or if China does, we are theoretically under this bill not to trade with them". (C-Span, 2010, 0:44)

Despite such a rational opinion, a military strike on Iran astonishingly remains a preferred tactic by certain individuals within the U.S.

government who fear that Iran will never bow to Western sanctions which in turn, have only consolidated the Iranians' national resolve and encouraged them to reach self-sufficiency in many economic areas, including gasoline production (Fars News Agency, 2011, ¶ 9). Those in favor of a military strike against Iran advocate a conventional air attack that would "only" target specific facilities in Iran's nuclear programs, though what fails to be mentioned is that almost all nuclear infrastructures are situated in or around populated areas (Cook & Roshandel, 2009, p. 119). Among these candidates is John Bolton, a former ambassador to the UN during the Bush administration and a potential presidential candidate of 2012, who insists that the U.S. needs to be more assertive and "bomb Iran" (Krieger, 2010, ¶ 1). He outlined this position further:

> "The most likely outcome with respect to Iran is that it gets nuclear weapons and very very soon...Given that diplomacy has failed, given that sanctions have failed, the only alternative to an Iran with nuclear weapons is a limited military strike against the nuclear weapons program". (Kreiger, 2010, ¶ 7)

Dismissing the argument that a strike would cause even more regional instability (or an array of blowback effects), Bolton points to documents provided by WikiLeaks which have revealed Arab support for a U.S. attack (Krieger, 2010, ¶ 7). Leaked cables have in truth revealed such support: the Saudi ambassador to the U.S., Adel A. Al-Jubeir, has implored Washington to "cut off the head of the snake" (Mather, 2010, ¶ 2). Abu Dhabi's crown prince, Mohammed bins Zayed al Nahyan, referred to Iran as an "existential threat" and referred to Ahmadinejad as "Hitler" (Mather, 2010, ¶ 2). King Hamad ibn Isa Al Khalifa of Bahrain told U.S. general David Petraeus, "that program must be stopped...The danger of letting it go on is greater than the danger of stopping it" (Mather, 2010, ¶ 3). The Kuwaiti interior minister Jaber Khaled al-Sabah believes "the U.S. will not be able to avoid a military conflict with Iran, if it is serious in its intention to prevent Tehran from achieving nuclear weapons" (Mather, 2010, ¶ 3). Chas Freeman, a journalist for the *New York Times*, suggests that such leaks do not necessarily make war with Iran more probable and if anything, has simply raised Iran's prestige by adding to the persistent estimation (or overestimation) of its influence and abilities as a regional power (2010, ¶ 2).

However, this argument is challenged when contrasted with recent U.S. public polls, which reveal alarming figures. The polls taken in 2010 disclose the highest public support for a military option: 65% to 25% in the Fox news/OD poll and 60% to 30% in the McLaughlin poll. According to the CNN/ORC poll taken in February 2010, 59%

supported military action and only 39% opposed it (Gilboa, 2010, p. 5). In the April-May 2010 Pew Global Attitudes Project, 64% of the respondents said that "it is more important: to prevent Iran from developing nuclear weapons, even if it means taking military action," than "to avoid a military conflict with Iran, even if it means they may develop nuclear weapons" (Gilboa, 2010, p. 5). Only 24% said that it is more important to "avoid a military conflict" (Gilboa, 2010, p. 4).

Iran's Response to U.S. Foreign Policy

Figure 2.3

Iranian President Mahmoud Ahmadinejad speaking at Columbia University on 24 September 2007. (Photo by Daniella Zalcman, provided under a Creative Commons License.)

Iran is in a win-win situation, according to Yassamine Mather, an Iranian socialist in exile for her leftist political activities. If the U.S. negotiates with the regime, Iran's regional prestige will strengthen (Mather, 2010, ¶ 13). On the other hand, if they continue to reinforce economic sanctions or opt for a military strike, the regime will likewise strengthen by destroying the large domestic opposition (Mather, 2010, ¶ 13). Thus far, Iran has largely ignored U.S. and UN Security Council sanctions and has responded with rhetorical counterattacks. For instance,

Ahmadinejad's senior advisor, Mojtaba Samereh Hashemi, has brushed aside any allegations of wounds in the Iranian economy, stating that there have been no "noticeable effects" (Erdbrink, 2010, ¶ 38). President Ahmadinejad himself has on numerous occasions mocked the ineffectiveness of sanctions and the potential for a military strike against Iran's nuclear facilities:

> "Who do you think is going to attack us? The Israeli regime?...We don't consider the regime in our equations, let alone attacking us. They say we'll issue sanctions? Okay do it. How many resolutions have you issued so far? Four? Make it 4,000". (The Jerusalem Post, 2010, ¶ 6-7).

Although such a response may not be the most effective idiom in the realm on negotiations, it is certainly efficient in highlighting certain truths—namely that sanctions generally do fail to achieve their intended aims. In an interview published by the Fars News Agency, Ahmadinejad equated the sanctions to a futile pursuit and pointed out that they have created new opportunities for progress: "The Iranian nation learned to rely on their (own) resources and capabilities...and as a result, made great scientific and industrial achievements" (Tehran Times, 2011, ¶ 3). As proof, he recited the successful implementation of the subsidy reform plan, the rise in foreign investment in Iran, and Iran's political and economic stability (Tehran Times, 2011, ¶ 4). In response to Obama's 08 March 2011 announcement of a one-year extension of sanctions, Foreign Minister Ali Akbar Salehi said in a radio interview with Iran's state news agency that U.S. efforts are an attempt to wage "psychological warfare" against the Islamic Republic with the intention of protecting its power and authority (United Press International [UPI], 2011, ¶ 2).

Under the umbrella of Obama's overtures to "the Muslim world," the U.S. president has directly stated to Iran: "We have serious differences that have grown over time. My administration is now committed to diplomacy that addresses the full range of issues before us, and to pursuing constructive ties among the United States, Iran and the international community" (Fox News, 2009, ¶ 3). This phrase implies that there exists willingness on the part of the U.S. to negotiate with Iranian leaders, which is failing to be reciprocated. Whether this is true or not, what is apparent is a lack of trust on both sides of the relationship. Ayatollah Ali Khamenei has publicly announced his mistrust of U.S. intentions and has urged Iranian representatives to be "extremely careful" when dealing with the U.S. (Erdbrink & Branigin, 2009, ¶ 8) for a move towards compromise would be seen as a sign of weakness and would encourage the U.S. to exert even greater pressure. Khamenei's rhetoric has revealed that the Obama administration is viewed by Iranian officials as a wolf in sheep's clothing:

> "Whenever they smile at the officials of the Islamic revolution, when we carefully look at the situation, we notice that they are hiding a dagger behind their back, they have not changed their intentions. What we have witnessed is completely the opposite of what they have been saying and claiming. On the face of things, they say, 'Let's negotiate.' But alongside this, they threaten us and say that if these negotiations do not achieve a desirable result, they will do this and that". (Quoted in Erdbrink & Branigin, 2009, ¶ 7-9)

Figure 2.4

U.S. President Barack Obama gives a special Nowruz video message to the people and leadership of Iran, 18 March 2009. (Photo by Chuck Kennedy, provided by Wikimedia Commons under a Creative Commons License.)

To a certain extent this may be true for sanctions against Iran have not been lifted and neither has the U.S. ruled out the possibility of a military strike.

Despite Iran's official rhetoric of invincibility, a U.S. military attack is a large concern for Iran which finds itself in an asymmetrical conflict: "Our armed forces will cut the hand of anyone in the world before it pulls the trigger against the Iranian nation" (Associated Press, 2009, ¶ 3). "No power dares imagine an invasion against Iran" and "The Iranian nation will resist all invaders" (Associated Press, 2009, ¶ 6). Although these remarks may appear threatening, one can recall Ansari's statement that Iranian hostility towards U.S. cannot be understood outside the political malaise that forms U.S.-Iran relations. It begs the question: is it not threatening when the U.S. claims that "no options are off the table"? So how does the issue of sanctions in regards to the nuclear program transcribe into public opinion?

The WPO asked Iranians in favour of the nuclear program whether they would "favour or oppose an agreement whereby the current sanctions against Iran would be removed and Iran would continue its nuclear energy program, except that it would agree not to enrich uranium?" (2010, p. 23). Among all Iranians, a 54% majority opposed such an agreement, with 31% in favor (WPO, 2010, p. 23). Those who opposed this agreement (or did not answer) were then asked: "Would you favor or oppose an agreement whereby the current sanctions against Iran would be removed and Iran would continue its uranium enrichment program, but would agree to grant international inspectors unrestricted access to all Iranian nuclear facilities to make sure that it is not making an atomic bomb" (WPO, 2010, p. 23). Fifty-two percent of those who had opposed the first question were in favor of the second (WPO, 2010, p. 23).

Conclusion

In reviewing the general rhetoric produced by Iran and U.S. officials, what becomes strikingly apparent is that a "Grand Bargain" to resolve all issues is non-existent. Keeping in mind that the nuclear issue is only a part of the whole, what negotiators of Iran's nuclear issues have ignored is the core issue: a deeply rooted lack of trust between both nations (Kahn, 2010, p. 107). Without sorting this problem out, any settlement of the nuclear dispute would be ineffective in the long run. Looking back on Obama's failed outreach to Iran, Foreign Service officer John Limbert eloquently states, "diplomatic efforts...foundered on mutual suspicion, political ineptitude, misreading signals, bad timing, and the power of inertia...Officials on both sides seemed unable to get beyond their classic responses" (Gerecht, 2010, ¶ 6).

The hope of beginning to build the foundations of a trusting relationship cannot thrive in a climate in which the official narrative of the other is one of hostility and blatant accusation. The "bad guy" narrative of Iran projected by the U.S. must stop and the victimization narrative projected by Iran must in turn, begin to diminish. An anti-Iran and anti-U.S. rhetoric, which as public polls suggest, manufactures public consent and perpetuates the status quo, has the potential to stimulate detrimental outcomes, such as warfare. Only once rhetoric changes can there begin a new phase whereby both nations will engage in genuine bilateral engagements. However, this is where both nations will be tested on their commitment in truly engaging in a constructive relationship. The U.S. must understand that Iran may be willing to renounce its nuclear program if the former makes adjustments in the course of its foreign policies pertaining not only to Iran but to the Middle East at large (Kahn, 2010, p. 106). Moreover, the U.S. must avoid having double standards in its foreign policy making: "what is fair for Israel should be fair for Iran or any other state in the Middle East" (Kahn, 2010, p. 106). Roger Howard, a writer and broadcaster on international relations, sums up these pointers:

> "[America] would need to fit the issue into a wider Middle East picture and find ways of making Iran feel less threatened. In return for cessation of uranium enrichment, or for more effective guarantees that it would not be used for a weapons program, Washington could offer not only to lift all sanctions but also to drop calls for a regime change and undertake not the meddle in Iran's domestic affairs; pull back its military presence in the regional and pressure Israel into surrendering or scaling down its nuclear arsenals". (Quoted in Kahn, 2010, p. 106)

The current Iranian hard-liner government's strategy can be summed by Ahmadinejad's statement: "If you pull back they will push ahead but it you stand against them, because of this resistance they will back off" (Kahn, 2010, p. 107). This strategy is counterproductive and must be altered for the U.S. has shown its determination to reinforce sanctions while the military option continues to loom in the shadows. Perhaps there will need to be a turn in Iranian governance, whereby decision makers would advocate a less confrontational policy pertaining to the U.S. There has been in the past with Akbar Hashemi Rafsanjani and Mohammad Khatami and there is no reason to doubt this will occur in the future. On a positive note and in the words of Saira Khan: "With...time and efforts and good intentions it is not impossible to terminate the 30-year old intractable conflict" (2009, p. 108).

References

Ahmadinejad, M. (2010). Address by H.E. Dr. Mahmoud Ahmadinejad, President of the Islamic Republic of Iran, Before the 65th Session of the United Nations General Assembly. New York: Permanent Mission of the Islamic Republic of Iran, September 23.
http://www.un.org/en/ga/65/meetings/generaldebate/Portals/1/statements/63 42085573815625001R_en.pdf

Ansari, A.M. (2006). *Confronting Iran: the Failure of American Foreign Policy and the Next Great Conflict in the Middle East.* New York, NY: Basic Books.

Associated Press. (2009). Ahmadinejad Warns Against Military Attack on Iran. *The Washington Times*, September 22.

http://www.washingtontimes.com/news/2009/sep/22/ahmadinejad-warns-against-military-attack-iran/?page=1

Bush, G.W. (2002). State of the Union Address, January 29. http://stateoftheunionaddress.org/2002-george-w-bush

Chubin, S. (2009). In J. D. Green, F. Wehrey, & C. Wolf Jr. (Eds.), *Understanding Iran* (pp. 52-64). Santa Monica, CA: RAND Corporation. http://www.rand.org/content/dam/rand/pubs/monographs/2008/RAND_MG 771.pdf

Constitution of Iran (1979). *Constitution of the Islamic Republic of Iran*. Bern: University of Bern. http://www.servat.unibe.ch/icl/ir00000_.html

Cook, A.H., & Roshandel, J. (2009). *The United States and Iran: Policy Challenges and Opportunities*. New York: Palgrave MacMillan.

Cordesman, A.H. & Seitz, A. (2009). *Iranian Weapons of Mass Destruction: the Birth of a Regional Nuclear Arms Race?* Washington, DC: Center for Strategic and International Studies.

C-Span. (2010). Ron Paul: Iran Sanction = Act of War [Video]. SpeakUpFightBack, April 23. http://www.youtube.com/watch?v=vIO-4v8qpYc

Dahl, F., & Westall, S. (2010) U.N Nuclear Assembly Rejects Arab Move Targeting Israel. *Reuters*, September 24. http://www.reuters.com/article/idUSTRE68N1O620100924

Delpech, T. (2006). *Iran and the Bomb: The Abdication of International Responsibility*. New York: Columbia University Press

Erdbrink, T. (2010). Transcript of interview with Ahmadinejad adviser Mojtaba Samareh Hashemi. *The Washington Post*, November 23. http://www.washingtonpost.com/wp-dyn/content/article/2010/11/23/AR2010112305058.html

Erdbrink, T. & Branigin, W. (2009). Iran's Supreme Leader Warns Against Negotiating with U.S. *The Washington Post*, November 4. http://www.washingtonpost.com/wp-dyn/content/article/2009/11/03/AR2009110301397.html

Erdbrink, T. & Kessler, G. (2010). Ahmadinejad Makes Nuclear Claims, Stifles Protest on Revolution's Anniversary. *The Washington Post*, February 12. http://www.washingtonpost.com/wp-dyn/content/article/2010/02/11/AR2010021100456.html?nav=emailpage

Fars News Agency (2011). US, Europe Continue Increasing Trade with Iran Despite Sanctions. *Fars News Agency*, April 2. http://english.farsnews.com/newstext.php?nn=9001134086

Fox News (2009). Obama Reaches Out to Iran, Looks for Engagement. *Fox News*, March 19. http://www.foxnews.com/politics/2009/03/19/obama-reaches-iran-looks-engagement/

Freeman, C. (2010). Why Iran Loves Wikileaks. *The New York Times*, December 4. http://www.nytimes.com/2010/12/05/opinion/05freeman.html

Gerecht, R.M. (2010). Diplomatic Illusions: Why Negotiations with Iran will Never Work. *The Weekly Standard*, December 13. http://www.weeklystandard.com/articles/diplomatic-illusions_520705.html

Gilboa, E. (2010) American Public Opinion Towards Iran's Nuclear Program: Moving Towards Confrontation. BESA Center Perspectives Papers,117, October 24. Ramat Gan, Israel: Begin-Sadat Center for Strategic Studies, Bar-Ilan University.
http://www.biu.ac.il/SOC/besa/docs/perspectives117.pdf

Green, J.D., Wehrey, F., & Wolf, C., Jr. (Eds.). (2009). Understanding Iran. Santa Monica, CA: RAND Corporation.
http://www.rand.org/content/dam/rand/pubs/monographs/2008/RAND_MG 771.pdf

Herman, E.S., & Chomsky, N. (1988) *Manufacturing Consent: the Political Economy of the Mass Media*. New York, NY: Pantheon Books.

Horovitz, D. (2010). Editor's Notes: Running Out of Time. *The Jerusalem Post*, November 26.
http://www.jpost.com/Opinion/Columnists/Article.aspx?id=196848

Hunter, R.E. (2010). Building Security in the Persian Gulf. Santa Monica, CA: RAND Corporation.
http://www.rand.org/pubs/monographs/2010/RAND_MG944.pdf

International Atomic Energy Agency Board of Governors [IAEA]. (2007). Implementation of the NPT Safeguards Agreement and relevant provisions of Security Council resolutions 1737 (2006) and 1747 (2007) in the Islamic Republic of Iran. Vienna: International Atomic Energy Agency, November 15.
http://www.iaea.org/Publications/Documents/Board/2007/gov2007-58.pdf

———— . (2009). Implementation of the NPT Safeguards Agreement and relevant provisions of Security Council resolutions 1737 (2006), 1747 (2007), 1803 (2008) and 1835 (2008) in the Islamic Republic of Iran. Vienna: International Atomic Energy Agency, June 5.
http://www.iaea.org/Publications/Documents/Board/2009/gov2009-35.pdf

The Jerusalem Post. (2010). Ahmadinejad wants Obama TV Debate. *The Jerusalem Post*, February 8.
http://www.jpost.com/IranianThreat/News/Article.aspx?id=183391

Kahn, S. (2010). *Iran and Nuclear Weapons: Protracted Conflict and Proliferation*. New York: Routledge.

Kaufman, S. (2011). Clinton: Iran Sanctions Working, Need to be Maintained. America Government Archive, January 10.
http://www.america.gov/st/peacesec-english/2011/January/20110110120608nehpets4.630679e-02.html

Kissinger, H. (1966). *American Foreign Policy: Three Essays*. New York: W.W. Norton.

Krieger, H.L. (2010) Bolton: Military strike only way to stop Iran nukes. *The Jerusalem Post*, November 30.
http://www.jpost.com/International/Article.aspx?id=197422

Mather Y. (2010). Lions, Foxes and Asses. *The Weekly Worker*, December 9.
http://www.cpgb.org.uk/worker2/index.php?action=viewarticle&article_id=1004 200

Mehr News Agency. (2005), Iran's Statement at IAEA Emergency Meeting. *Mehr News Agency*, August 10.
http://www.fas.org/nuke/guide/iran/nuke/mehr080905.html

Non-Proliferation of Nuclear Weapons (NPT). (2000). The Treaty on the Non-

Proliferation of Nuclear Weapons. New York: Department for Disarmament Affairs, United Nations.

http://www.un.org/en/conf/npt/2005/npttreaty.html

Obama, B. (2011). Notice on the Continuation of the National Emergency with Respect to Iran. Washington DC: White House, Office of the Press Secretary, March 08.

http://www.whitehouse.gov/the-press-office/2011/03/08/notice-continuation-national-emergency-respect-iran

Obama, B., Sarkozy, N., & Brown, G. (2009). Statements by President Obama, French President Sarkozy, and British Prime Minister Brown on Iranian Nuclear Facility. Washington, DC: White House, Office of the Press Secretary, September 25.

http://www.whitehouse.gov/the_press_office/Statements-By-President-Obama-French-President-Sarkozy-And-British-Prime-Minister-Brown-On-Iranian-Nuclear-Facility/

Press TV. (2010). Iranian President Mahmoud Ahmadinejad: "Iran's Nuclear Inalienable Right Non-Negotiable" [Video]. Multionym, November 20.

http://www.youtube.com/watch?v=Wwh7unPzsnY

Sadjadpour, K. (2009). Reading Khamenei: The World View of Iran's Most Powerful Leader. Washington, DC: Carnegie Endowment for International Peace.

Shanker, T. (2011). Warning Against Wars Like Iraq and Afghanistan. The New York Times, February 25.

http://www.nytimes.com/2011/02/26/world/26gates.html

Stewart, P. (2010). Gates Sees Iran Rift, Says Strike Would Unite Country. Reuters, November 16.

http://www.reuters.com/article/idUSTRE6AF3G720101116

Tehran Times. (2011). Western Companies Circumventing Iran Sanctions: Ahmadinejad. Tehran Times, March 7.

http://www.tehrantimes.com/index_View.asp?code=236939

UPI. (2011). Iran Snubs U.S. Sanctions Extension. United Press International, March 23.

http://www.upi.com/Top_News/World-News/2011/03/23/Iran-snubs-US-sanctions-extension/UPI-55821300915984/

U.S. Energy Information Administration (EIA). (2010). Iran Analysis. Washington, DC: U.S. Department of Energy.

http://www.eia.doe.gov/countries/cab.cfm?fips=IR

White, J.B. (1984). When Words Lose Their Meaning: Constitutions and Reconstitutions of Language, Character and Community. Chicago: University of Chicago Press.

World Public Opinion (WPO). (2010). An Analysis of Multiple Polls of the Iranian Public. College Park, MD: The Program on International Policy Attitudes at the University of Maryland.

CHAPTER THREE

HARD POWER AND IRAQ:
DESTABILIZATION, INVASION AND OCCUPATION

Corey Seaton

After World War II the U.S. helped to found many of the multi-lateral organizations, partly in an effort to keep the Soviet Union and Communism from spreading throughout the world. However, since the fall of the USSR, U.S. power has been unrivalled and it has enjoyed a world where what it wants is deemed what is right for the entire world. To this end, the Americans have employed a strategy that ensures a continually unipolar world where the U.S. has no peer competitor (Chomsky, 2004, p. 11). The means by which they have accomplished this has been through the threat or use of force, backed by one of the largest and most advanced militaries that the Earth has ever seen, and with one of the largest economies in the world. This chapter thus focuses on the preeminent feature of U.S. foreign policy: the deployment of what some now call "hard power".

Global dominance has made the U.S. arrogant in its dealings with foreign powers and this can be seen in its growing disdain for the very treaties and multilateral organizations that the U.S. helped to establish, as well as international public opinion, while relying on its military and economic power to coerce others to bend to U.S. interests and objectives (Zakaria, 2008, p. 222). Unaccustomed to this new dynamic of being the sole superpower and needing to test their reach and show the world the power of the U.S., it would impose its vision of order on the world, whether the world liked it or not.

The grand strategy for enforcing American power relies on the idea of pre-emptive war against targets that are deemed to be threats to American power and relying on force to maintain U.S. hegemony when the need arises (Chomsky, 2004, p. 16). This doctrine of pre-emptive war is very troubling, especially when taking into account the fact that the perception of what is a threat is highly subjective, sometimes deliberately fanciful. Of course, if this doctrine were to be applied evenly to all those who might pose a threat to American interests, they would be invading countries all the time. As this is not the case, countries that that are targets for pre-emptive war have to have several characteristics: a) they must be virtually defenseless; b) they must be important enough to be worth the trouble; and, c) they must be able to be portrayed as the ultimate evil and threat to U.S. survival (Chomsky, 2004, p. 17).

This is called "hard power" and, borrowing from Joseph Nye's conceptualization, it "rests on an actor's capacity to get others to change their position through either the making of threats or the proffering of incentives" (Lock, 2010, p. 33). Yet, unlike Nye, my attempt here is to show U.S. power in its relations to specific others, as with Lock (2010, p. 46):

> "Firstly, to equate power with certain resources is to perpetuate the 'vehicle fallacy' that scholars (including Nye) have repeatedly warned against....The problem here is obvious; actors with great power resources do not always exercise power over others in terms of getting them to do what it is that they want them to do. Secondly, the assumption that power is possessed by one actor implies that we can successfully evaluate the exercise of power by the U.S. simply by examining what it is the US does and without even considering those over whom American power is exercised".

Hard power can manifest in three different ways: military intervention, sanctions, and aid. Military intervention involves the deliberate act of a nation or a group of nations to introduce military forces (perhaps into the course of an existing controversy); sanctions are trade penalties imposed on a country; and, aid is simply a large bribe. While they differ in how they are applied, the end result remains the same: to coerce others to follow the lead of the strong and the leader of the strong in today's world is the U.S. American leaders claim to be doing this to bring peace to a given region and to protect human rights. However, the way that these forms of power are applied can be counter-productive and detrimental to both peace and human rights.

By analyzing how military intervention and economic sanctions are applied we can see their effects on the international stage and observe how disruptive they are to international actors. In particular, I will be looking at the invasion of Kuwait by Iraq and how the international community and in particular the U.S. responded to the situation and punished Iraq, and the invasion of Iraq and its after effects on the entire region.

UN Sanctions against Iraq and Desert Storm

The Iraqi invasion of Kuwait began on 02 August 1990. The reason for the invasion was a dispute over increasing oil production (and thus falling revenues for Iraq), the boundaries of the Rumaila oilfields (Hayes, 1990, ¶ 1), and a debt accrued over the course of the Iran-Iraq War (Reuters, 1990, ¶ 1). The UN responded to the crisis by passing resolutions with the goal of convincing Saddam Hussein to stand down and halt Iraq's annexation of Kuwait. However, Hussein refused to comply and the U.S. and its allies intervened. The air campaign started on 16 January 1991 with the goal of hindering Iraq's armed forces, supply lines, and command (Gellman, 1991). Once the aerial bombardment was over, the ground assault started and while the Iraqi armed forces offered some resistance ultimately they were easily overrun and Kuwait was liberated from its Iraqi captors. The Gulf War was a defining moment for the U.S. as it was the first step in showing the world that it was now the leading power and that it would do whatever it takes to maintain that hold on power. Iraq was only the beginning in establishing the norm of preventative war and can be seen as an experiment (Chomsky, 2004, p. 21). However, we must also consider the devastation suffered by the Iraqi population.

The bombing campaign lasted 43 days and while we were told that coalition forces would only target Iraq's offensive capabilities, this simply is not true. *The Washington Post*'s Barton Gellman (1991) found that the targets and purpose of the air campaign were much broader than initially claimed. Initially, primary targets were chosen so as to minimize both civilian and military casualties alike. However, the bombings targeted not only military targets, but also civilian infrastructure such as: oil refineries, electrical plants, water treatment and, sewage facilities. These installations were being attacked not because they would directly impact the outcome of the war; they were attacked because by targeting Iraq's infrastructure this would create leverage that would be used after the war (Gellman, 1991, p. 1). This was done deliberately, especially near the end of the war, so as to accelerate the effects of the economic sanctions and to then use devastation as leverage against Saddam Hussein after the war (Gellman, 1991, p. 2). Knowing that the Iraqis could not repair their facilities without help from Western powers, they crippled Iraq's ability to support itself in the hopes of gaining an advantage over Saddam Hussein forcing him to comply with the resolutions placed upon him. The bombings in and of themselves crippled Iraq and left the country in tatters, but in conjunction with the severe sanctions that were placed on Iraq was tantamount to biological warfare (Chomsky, 2008a, ¶ 46).

Figure 3.1

The Highway of Death. Civilian and Iraqi military vehicles litter a section of the Kuwait City Highway attacked by Allied aircraft during Operation Desert Storm. 28 February 1991. (Source: Capt. R.J. Worsley, released to the public, and provided by Wikimedia Commons under a Creative Commons License.)

United Nations Security Council Resolution 687, passed on 03 April 1991, specified:

> "Effective immediately, that the prohibitions against the sale or supply to Iraq of commodities or products, other than medicine and health supplies, and prohibitions against financial transactions related thereto contained in resolution 661 (1990) shall not apply to foodstuffs notified to the Security Council Committee established by resolution 661 (1990) concerning the situation between Iraq and Kuwait or, with the approval of that Committee, under the simplified and accelerated 'no-objection' procedure, to materials and supplies for essential civilian needs as identified in the report of the Secretary-General dated 20 March 1991, and in any further findings of humanitarian need by the Committee". (UN, 1991, Section F, Art. 20).

Evidence has since surfaced, in documents from the Defense Intelligence Agency (DIA), that "the U.S. government intentionally used sanctions against Iraq to degrade the country's water supply after the Gulf War" (Nagy, 2001, ¶ 1). This is contrary to the Geneva Convention. Article 54(2) of the "Protocol Additional to the Geneva Conventions of 12 August 1949, and relating to the Protection of Victims of International Armed Conflicts (Protocol 1)" states:

"2. It is prohibited to attack, destroy, remove or render useless objects indispensable to the survival of the civilian population, such as foodstuffs, agricultural areas for the production of foodstuffs, crops, livestock, drinking water installations and supplies and irrigation works, for the specific purpose of denying them for their sustenance value to the civilian population or to the adverse Party, whatever the motive, whether in order to starve out civilians, to cause them to move away, or for any other motive".

The primary DIA document obtained by Thomas Nagy titled, "Iraq Water Treatment Vulnerabilities," dated 22 January 1991, spelled out how sanctions would prevent Iraq from supplying clean water to its citizens:

"Iraq depends on importing specialized equipment and some chemicals to purify its water supply, most of which is heavily mineralized and frequently brackish to saline. With no domestic sources of both water treatment replacement parts and some essential chemicals, Iraq will continue attempts to circumvent United Nations Sanctions to import these vital commodities. Failing to secure supplies will result in a shortage of pure drinking water for much of the population. This could lead to increased incidences, if not epidemics, of disease. [1]

"Given the high level of pollutants and bacteria in the Iraqi water system, the DIA predicted that if left untreated—as was the U.S. plan— "epidemics of such diseases as cholera, hepatitis, and typhoid could occur". (Quoted in Nagy, 2001, ¶ 4, 5).

In another DIA document, titled "Disease Information" (also dated 22 January 1991) states the following under, "Subject: Effects of Bombing on Disease Occurrence in Baghdad":

"Increased incidence of diseases will be attributable to degradation of normal preventive medicine, waste disposal, water purification/distribution, electricity, and decreased ability to control disease outbreaks. Any urban area in Iraq that has received infrastructure damage will have similar problems". (Quoted in Nagy, 2001, ¶ 16).

The DIA repeatedly notes in that document that "particularly children" will be affected by the outbreak of disease (Nagy, 2001, ¶ 17). As Nagy noted, "for more than ten years [since the Gulf War], the United States has deliberately pursued a policy of destroying the water treatment system of Iraq, knowing full well the cost in Iraqi lives. The United Nations has estimated that more than 500,000 Iraqi children have died as a result of sanctions, and that 5,000 Iraqi children continue to die every month for this reason" (2001, ¶ 39). As the documents show, the U.S. knew exactly what the impacts of the sanctions would be and deliberately chose to aggravate the consequences with full knowledge of the likely—and then resulting—impact on innocent civilians.

Instead of any sense of guilt or doubt, we have what one Pentagon official told Gellman (1991, p. 5):

> "People say, 'You didn't recognize that it was going to have an effect on water or sewage.' Well, what were we trying to do with sanctions—help out the Iraqi people? No. What we were doing with the attacks on infrastructure was to accelerate the effect of the sanctions".

Col. John A. Warden III, deputy director of strategy, doctrine and plans for the Air Force, agreed and told Gellman (1991, p. 5) that one purpose of destroying Iraq's electrical grid was that "you have imposed a long-term problem on the leadership that it has to deal with sometime". Warden added:

> "Saddam Hussein cannot restore his own electricity. He needs help. If there are political objectives that the U.N. coalition has, it can say, 'Saddam, when you agree to do these things, we will allow people to come in and fix your electricity.' It gives us long-term leverage". (Quoted in Gellman, 1991, p. 5)

And another U.S. Air Force planner told Gellman:

> "Big picture, we wanted to let people know, 'Get rid of this guy and we'll be more than happy to assist in rebuilding. We're not going to tolerate Saddam Hussein or his regime. Fix that, and we'll fix your electricity'". (1991, p. 5)

In other words, the U.S. clearly inflicted collective punishment, of a lethal and illegal nature (not to mention immoral) that held ordinary Iraqis hostage to the dictates of U.S. foreign policy. One can be forgiven for thinking that this policy was cold blooded—some may recall the famous interview on CBS' 60 *Minutes* between journalist Lesley Stahl and Madeleine Albright, U.S. Secretary of State in the Clinton administration, on 12 May 1996:

> *Lesley Stahl on U.S. sanctions against Iraq:* "We have heard that a half million children have died. I mean, that's more children than died in Hiroshima. And, you know, is the price worth it?"

> *Secretary of State Madeleine Albright:* "I think this is a very hard choice, but the price—we think the price is worth it".[2]

Iraq was blockaded by sea, air, and ground with no countries able to legally do business with Iraq. These were some of the most comprehensive sanctions the UN had ever applied against any nation. The reason for this was that the sanctions were aimed at keeping funding away from Saddam Hussein to prevent his purchase or development of any "weapons of mass destruction," and to coerce regime change from within and reshape Iraq to the needs of dominant powers on the Security Council. Furthermore, Iraq not only had to be repay its debt to Kuwait accrued prior to its invasion of Kuwait, but it also had to compensate Kuwait for the damages and items stolen during the invasion (UN, 1991, Section E, Art. 16). All of this was for naught as the sanctions failed in their objective of getting Iraq to pull out of Kuwait and the U.S. led military intervention in the conflict. While the sanctions were effective in choking Iraq of both money and supplies, they had little effect on Saddam Hussein as he continued to evade UN resolutions on disarmament and disclosing full capabilities.

The sanctions proved ineffective against Saddam Hussein and the people who were most affected were the Iraqi people. Since their implementation, the Iraqi population suffered tremendously because of the sanctions in conjunction and the war. Sanctions led to inflation rising into the triple digits and wages decreasing, which then led to people not being able to purchase food on the free market and forcing them to instead rely upon the black market economy, as well as forcing people to rely on Saddam Hussein's regime in order to survive, thereby strengthening the regime's hold (Chomsky, 2004, p. 141).

Rebuilding efforts were sporadic and temporary at best. The parts for this re-construction could not be purchased and Iraqis resorted to cannibalizing factories to do any repairs. Any factories that were not damaged were forced to either shut down or scale back production drastically and the lack of raw materials and spare parts led to widespread unemployment. Also, the destruction of water-treatment facilities and sanitation facilities led to the contamination of the nation's only sources of drinking water (Lynch, 1997, "Effect on the Environment"). Cities became veritable garbage dumps which posed a risk health for all, especially the poor, as they searched for food. This combined with the fact that sewage and water treatment plants were destroyed made a horrible situation into what is essentially biological warfare (Chomsky, 2008a, ¶ 44). Also, there were reports of aid, such as water trucks and medicines, which were not allowed to be sent to Iraq because they were so-called dual-use items that allegedly could be used to manufacture weapons of mass destruction, despite the fact that European weapons experts stated that it was impossible for these items to be used in that context (Chomsky, 2004, p. 128).

Regrettably, the segment of society which has been the most adversely affected by the sanctions was Iraq's sick and children. Since the imposition of the sanctions, the infant mortality rate increased significantly. Although, medical equipment and drugs were exempted from the sanctions there was little money to buy them, which led to constant shortages of drugs and medical products. Further, this shortage led to the interruption or complete lack of treatment for sick patients and is essentially a death sentence for these patients, especially for children with leukemia (Lynch, 1997, "Effects of Sanctions on Human Rights"). In fact, all of Iraq's children were at risk because of the deteriorating living conditions and hospitals, to the point where diseases once eradicated were returning. Further, from 1993-2003 there was a dramatic increase of mortality rates for children under five, which rose from 56 per 1,000 to 131 per 1,000 (Chomsky, 2008a, ¶ 47). This has led to the deaths of over 1.5 million Iraqis, many of whom were children, over the course of both the sanctions and the Gulf War (Halliday, 2000, ¶ 13).

However, the UN was not completely blind to suffering of the Iraqi people and in 1996 the UN instituted a program that would allow for Iraq to sell oil and use those funds to procure much needed supplies, medicine, and food. In an interview with Charlie Rose, Paul Volcker, chairman of the UN's independent inquiry committee on the Oil-for-Food program, described the program as "designed to sustain the Iraqi population nutritionally and medical wise" (Rose, 2004, 21:11). It did help to alleviate some of the shortages that the Iraqi population was facing such as food and drug shortages and a deteriorating medical system. While the oil-for-food program did some good, it did not halt the collapse of the healthcare system and the deterioration of access to clean water (Chomsky, 2004, p. 128).

Some of the shortcomings of the oil-for-food program were built in. For instance, the UN took 35% of the gross sales of oil and put 30% in a fund to compensate Kuwait for Iraq's seven months of occupation. The situation was so dire that Denis Halliday and Hans Von Sponeck, both distinguished international diplomats, resigned their posts as directors of the Oil-for-Food Program because they saw the program as "genocidal" toward the Iraqi people (Chomksy, 2008a, ¶ 46). In addition, they both resigned their posts because they had been barred from speaking to the U.S. media, and the Clinton administration prevented Von Sponeck from addressing the Security Council about the effects of the sanctions (Chomsky, 2008a, ¶ 47). They spoke out about the suffering of the Iraqi people, but it was largely ignored by the U.S. media.

Furthermore, not only was the program inadequate to meet the needs of the Iraqi people, it was also mismanaged and corrupt. For instance, Volker states that many international corporations that bought oil or supplied aid were able to give large kickbacks to not only Saddam Hussein, but to the man in charge of the program. Saddam Hussein was able to smuggle oil out of the country, which was against the sanctions, and when the Security Council found out they did nothing and allowed it to continue (Rose, 2004, 24:10).

Figure 3.2

Near Al Najaf, Iraq, 08 April 2003–U.S. military food distribution and crowd control. From the original military caption: "U.S. Army military police provide crowd control while Iraqi citizens line up for food and water being distributed to citizens in need. The U.S. military is working with international relief organizations to help provide food and medicine for the Iraqi people in support of Operation Iraqi Freedom. Operation Iraqi Freedom is the multinational coalition effort to liberate the Iraqi people, eliminate Iraq's weapons of mass destruction and end the regime of Saddam Hussein". (Source: U.S. Navy photo by Photographer's Mate 1st Class Arlo K. Abrahamson, released to the public and provided by Wikimedia Commons under a Creative Commons License.)

In the end, the sanctions achieved the opposite of their intended effect. The ultimate goals of the sanctions were to oust Saddam Hussein from power and place another leader, one who could be controlled by the U.S. Instead of leading to rebellions against the tyrant, the sanctions crippled the middle class and impoverished the majority of the population, which allowed Hussein to consolidate his power and gain an even tighter control over the Iraqi people (Chomsky, 2004, p. 140-141). With no other alternatives people relied on Hussein to survive. This allowed Hussein to maintain power and quash any dissent that would jeopardize his control of Iraq. The only thing that the sanctions accomplished was punishing the Iraqi people for not ousting Hussein themselves. Iraq was left in tatters for over a decade and its people have suffered tremendously for the transgression of its leader. For the people of Iraq, the sanctions and Gulf War had been devastating, but after the events of 9/11 they could not have imagined it could get any worse. They were wrong; it was only the beginning of the U.S. pursuit of power and resources.

Justification for War?

After the attacks on 11 September 2001, the U.S. set for itself the mission to stop terrorism, at any cost.[3] Initially the U.S. invaded Afghanistan to chase Al Qaeda and make them pay for attacking the U.S. However, the U.S. leadership had also begun to fabricate Iraqi threats and make false assertions about Hussein's involvement with Al Qaeda (see Russert, 2003; Kessler & VandeHei, 2004). The media was used to spread lies about Hussein's Involvement and incite fear among the American people, so as to justify the invasion of Iraq.

In connection with "9/11," this is what Vice-President Dick Cheney had to say on NBC's *Meet the Press* in 2003 (Russert, 2003, ¶ 19-26):

> "MR. RUSSERT: The Washington Post asked the American people about Saddam Hussein, and this is what they said: 69 percent said he was involved in the September 11 attacks. Are you surprised by that?

> "VICE PRES. CHENEY: No. I think it's not surprising that people make that connection.

> "MR. RUSSERT: But is there a connection?

> "VICE PRES. CHENEY: We don't know. You and I talked about this two years ago. I can remember you asking me this question just a few

days after the original attack. At the time I said no, we didn't have any evidence of that. Subsequent to that, we've learned a couple of things. We learned more and more that there was a relationship between Iraq and al-Qaeda that stretched back through most of the decade of the '90s, that it involved training, for example, on BW and CW, that al-Qaeda sent personnel to Baghdad to get trained on the systems that are involved. The Iraqis providing bomb-making expertise and advice to the al-Qaeda organization.

"We know, for example, in connection with the original World Trade Center bombing in '93 that one of the bombers was Iraqi, returned to Iraq after the attack of '93. And we've learned subsequent to that, since we went into Baghdad and got into the intelligence files, that this individual probably also received financing from the Iraqi government as well as safe haven.

"Now, is there a connection between the Iraqi government and the original World Trade Center bombing in '93? We know, as I say, that one of the perpetrators of that act did, in fact, receive support from the Iraqi government after the fact. With respect to 9/11, of course, we've had the story that's been public out there. The Czechs alleged that Mohamed Atta, the lead attacker, met in Prague with a senior Iraqi intelligence official five months before the attack, but we've never been able to develop anymore of that yet either in terms of confirming it or discrediting it. We just don't know.

"MR. RUSSERT: We could establish a direct link between the hijackers of September 11 and Saudi Arabia.

"VICE PRES. CHENEY: We know that many of the attackers were Saudi. There was also an Egyptian in the bunch. It doesn't mean those governments had anything to do with that attack. That's a different proposition than saying the Iraqi government and the Iraqi intelligent service has a relationship with al-Qaeda that developed throughout the decade of the '90s. That was clearly official policy". [Emphases added]

In addition it was also falsely asserted that Iraq had "weapons of mass destruction" and would use them against the U.S. and its allies, and that Iraq was likely planning an attack on American soil (see Powell, 2003a). The George W. Bush administration also claimed Saddam Hussein was an immediate threat and must be preemptively removed as a matter of self-defence. By linking him with the trauma of 9/11, the U.S. had convinced the people, with blatant lies and deception, of the continuing threat Iraq posed to national security and American values (Chomsky, 2004, pp. 18-19).

Here are just a few examples of some of the assertions made by U.S. Secretary of State Colin Powell at the UN Security Council in 2003 (Powell, 2003b):

"There can be no doubt that Saddam Hussein has biological weapons and the capability to rapidly produce more, many more. And he has the ability to dispense these lethal poisons and diseases in ways that can cause massive death and destruction. If biological weapons seem too terrible to contemplate, chemical weapons are equally chilling. (¶ 6)

"Our conservative estimate is that Iraq today has a stockpile of between 100 and 500 tons of chemical weapons agent. That is enough agent to fill 16,000 battlefield rockets. (¶ 40)

"Saddam Hussein has chemical weapons. Saddam Hussein has used such weapons. And Saddam Hussein has no compunction about using them again, against his neighbours and against his own people. (¶ 43)

"Let me turn now to nuclear weapons. We have no indication that Saddam Hussein has ever abandoned his nuclear weapons programme. (¶ 47)

"On the contrary, we have more than a decade of proof that he remains determined to acquire nuclear weapons. (¶ 48)

"Saddam Hussein is determined to get his hands on a nuclear bomb. (¶ 54)

"Let me talk now about the systems Iraq is developing to deliver weapons of mass destruction, in particular Iraq's ballistic missiles and unmanned aerial vehicles, UAVs. (¶ 64)

"Saddam Hussein's intentions have never changed. He is not developing the missiles for self-defense. These are missiles that Iraq wants in order to project power, to threaten, and to deliver chemical, biological and, if we let him, nuclear warheads. (¶ 73)

"Now, unmanned aerial vehicles, UAVs. (¶ 74)

"Iraq has been working on a variety of UAVs for more than a decade. This is just illustrative of what a UAV would look like. This effort has included attempts to modify for unmanned flight the MiG-21 and with greater success an aircraft called the L-29. However, Iraq is now concentrating not on these airplanes, but on developing and testing smaller UAVs, such as this. (¶ 75)

"UAVs are well suited for dispensing chemical and biological weapons". (¶ 76)

And this, from Powell's concluding statement to the UN (Powell, 2003c, ¶ 15, 16):

"We know that Saddam Hussein is determined to keep his weapons of mass destruction; he's determined to make more. Given Saddam Hussein's history of aggression, given what we know of his grandiose plans, given what we know of his terrorist associations and given his determination to exact revenge on those who oppose him, should we take the risk that he will not some day use these weapons at a time and the place and in the manner of his choosing at a time when the world is in a much weaker position to respond?

"The United States will not and cannot run that risk to the American people. Leaving Saddam Hussein in possession of weapons of mass destruction for a few more months or years is not an option, not in a post-September 11th world".

However, the international community knew that Iraq had no weapons and they knew this because the UN had sent a group of weapons inspectors to investigate the issue. Hans Blix was the leader of a group of over 200 inspectors and support staff from a combination of 60 countries. They conducted a search for weapons of mass destruction and did a thorough sweep with over 400 inspections of 300 facilities with the capability to manufacture chemical weapons (Blix, 2003, ¶ 5). These buildings were checked and re- rechecked for contents new and old from the previous weapons checks in 1998, and did so with no advance notice and received prompt access from Iraqi personnel. They even used ground radar to detect for any buried equipment (Blix, 2003, ¶ 6). What they found and took was over 200 chemical and more than 100 biological samples, and concluded that their results are consistent with what Iraq had declared (Blix, 2003, ¶ 8). They concluded that Iraq possessed no such weapons of mass destruction and that they did not possess any of the weapons they claimed to own.

The U.S. government claimed that it had intelligence stating that Iraq had these weapons and were hiding them, and claimed links to Osama Bin Laden and Al Qaeda, despite the fact that there were no such links (Chomsky, 2008. p. 6). People within the intelligence community in the U.S. and worldwide saw no credibility in the claims of the Bush administration and believed that they opposed inspection, because they feared nothing would be found (Chomsky, 2004, p. 18). It was all just a smokescreen to persuade the American people by using their fear to support a war (Chomsky, 2004, p. 18). Even with of all this concocted evidence many people in America did not want to invade Iraq, but they were not to be listened to as it was "in the national interest" to start a war. The reason is simple: oil. Iraq has one of the largest oil reserves in the world and having military bases at the heart of the world's energy supply would leave the U.S. in control of the world's energy (Chomsky, 2008, ¶ 6).

The Iraq War and the Destabilization of a Region

The U.S. Invasion of Iraq started off as a disaster, both politically and diplomatically from the beginning. The primary reason for going to war was to disarm Iraq of its alleged "weapons of mass destruction," but from the outset more ambiguous aims were added. In his speech to the American people, then President George W. Bush announced as the invasion of Iraq began on 19 March 2003: "at this hour American and coalition forces are in the early stages of military operations to disarm Iraq, to free its people and to defend the world from grave danger" (CNN, 2003). The U.S. decided to invade and defend the world from the future acts of terrorism that Saddam Hussein could potentially commit and to do their own search of Iraq for WMDs.

The U.S. invasion was, however, done without the approval of the UN Security Council and went against global public opinion. In an interview with Owen Bennett-Jones (2004), Kofi Annan, then Secretary General of the UN, clearly stated that the invasion of Iraq was illegal; it violated the UN Charter, and did not conform to any mandate of the UN (¶ 10).

In fact there were multiple legal opinions and admissions that the invasion was illegal, from within the high ranks of government in the U.S. and the UK which joined in leading the invasion. Richard Perle, a key member of the Bush's Defense Policy Board which advised Donald Rumsfeld, told an audience in London back in November 2003: "I think in this case international law stood in the way of doing the right thing" (Burkeman & Borger, 2003, ¶ 1). From early on, he conceded that the war was illegal. He added, "international law...would have required us to leave Saddam Hussein alone" and that there was "no practical mechanism consistent with the rules of the UN for dealing with Saddam Hussein" (Burkeman & Borger, 2003, ¶ 3, 4).

In a "secret and personal" letter from Jack Straw (the UK Foreign Secretary in 2002) to Prime Minister Tony Blair, he "warned the prime minister that the case for military action in Iraq was of dubious legality;" Straw also stated that "regime change per se is no justification for military action" and "the weight of legal advice here is that a fresh [UN] mandate may well be required" (Smith, 2010, ¶ 1, 7). Also in the UK, Lord Bingham, a former Lord Chief Justice, explained that the British decision to invade Iraq along with the U.S. was "fundamentally flawed" in terms of its legality (BBC, 2008, ¶ 1). Again in the UK, in a minute dated 18 March 2003 from Elizabeth Wilmshurst (Deputy Legal Adviser) to Michael Wood (The Legal Adviser), copied to the Private Secretary, the Private Secretary to the Permanent Under-Secretary, Alan Charlton (Director Personnel) and Andrew Patrick (Press Office), Wilmshurst stated:

> "I regret that I cannot agree that it is lawful to use force against Iraq without a second Security Council resolution I cannot in conscience go along with advice—within the Office or to the public or Parliament—which asserts the legitimacy of military action without such a resolution, particularly since an unlawful use of force on such a scale amounts to the crime of aggression; nor can I agree with such action in circumstances which are so detrimental to the international order and the rule of law". (Wilmhurst, 2005, ¶ 3)

Wilmshurst resigned in March 2003 because she did not believe the war with Iraq was legal.

In The Netherlands, a partner in the invasion of Iraq, a high level inquiry, "in a damning series of findings on the decision of the Dutch government to support Tony Blair and George Bush in the strategy of regime change in Iraq...found the action had 'no basis in international law'" (Hirsch, 2010, ¶ 2). Willibrord Davids, a Dutch Supreme Court judge, said U.N. resolutions in the 1990s prior to the 2003 invasion have no authority for the invasion. In the 551-page report (Commisie van onderzoek besluitvorming Irak, 2010), the inquiry stated: "The Dutch government lent its political support to a war whose purpose was not consistent with Dutch government policy. The military action had no sound mandate in international law" (Hirsch, 2010, ¶ 3).

The International Commission of Jurists (ICJ) on 18 March 2003 also expressed its,

> "deep dismay that a small number of states are poised to launch an outright illegal invasion of Iraq, which amounts to a war of aggression. The United States, the United Kingdom and Spain have signalled their intent to use force in Iraq in spite of the absence of a Security Council Resolution. There is no other plausible legal basis for this

attack. In the absence of such Security Council authorisation, no country may use force against another country, except in self-defence against an armed attack". (ICJ, 2003, ¶ 1)

Even though the Security Council was against the invasion, the U.S. stated it did not need the permission of the UN to defend U.S. national security and that the UN was now irrelevant because it had not lived up their responsibility, which apparently consists of following orders from Washington (Chomsky, 2004, pp. 32-33). This shows us that the Bush administration had designs to invade Iraq no matter what. In the aftermath of the invasion the military began to search for WMDs, but to no one's surprise, none were found. So as to legitimate their invasion they stated the mission was to now fight terrorists in their own homes; otherwise, they would be in the streets of America.

The Fracture of Iraq under U.S. Occupation

Most insurgents in Iraq had no plans or even hopes of one day invading America and making them pay; they only wished to fight the people who they see as exploiting their country for its wealth (Zunes, 2007, p. 59). The truth of the matter is that the only reason there is a problem in Iraq with insurgency and terrorism is because of the invasion and occupation of Iraq; it is only a small percentage of groups that have aims of targeting the U.S. at home (Zunes, 2007, p. 60). All of this resentment and resistance stemmed from the policies that alienated segments of Iraq and the perception that the U.S. was trying to steal Iraq's wealth (Zunes, 2006, ¶ 1).

With the U.S. invasion being illegitimate, against worldwide opinion, and their search for WMDs fruitless, the U.S. now had to deal with a country that was held together by Saddam Hussein, who was more than just a dictator, but also a source of stability that ensured the coherence of Iraq. However, with Saddam Hussein removed from power that stability was gone, leaving people unsure of the future and looking for protection and security—all of which the invasion clearly would not provide. This inability or unwillingness to aid the people of Iraq has fostered mush resentment and resistance to the U.S. and the government they have installed (Zunes, 2007, p. 64). This is not hard to believe when one of the first decisions Washington made was to abolish the army and government bureaucracy, both of which were secular, thereby leaving a huge power vacuum and forcing people to seek protection from extremist sectarian parties and militias (Zunes, 2006, ¶ 8). The Shia community was able to better organize and mobilize when compared to the Sunni and secular groupings (Zunes, 2006, ¶ 12).When elections were held in 2004, it came as no surprise that Shia religious parties rose to power and marginalized other groups, while imposing a repressive form of Islamic law (Zunes, 2006, ¶ 12). As a result, political authority has been split along sectarian lines, with every decision being debated in terms of which group it could benefit or harm (Zunes, 2006, ¶ 9). This left the Iraqi government unstable, split all aspects of the government and military down sectarian lines, thus leaving the Iraqi government under the control and influence of the U.S.

Figure 3.3

An Iraqi woman looks on as U.S. Army Soldiers from 1st Battalion, 23rd Infantry Regiment, 3rd Stryker Brigade Combat Team search the courtyard of her house during a cordon and search in Ameriyah, Iraq. 14 May 2007. (Source: U.S. Army photo by Sgt. Tierney Nowland, dodmedia.osd.mil, in the public domain and provided by Wikimedia Commons under a Creative Commons License.)

This has led to a dramatic increase in violence. In particular, the use of suicide bombs increased dramatically with the targets of these attacks being U.S. military personnel and Iraqi civilians. Since the start of the U.S. occupation, the amount of suicide attacks in Iraq increased dramatically. The apparent goal was to overthrow the interim government of Iraq, not allow Iraqi forces to become effective, and drive U.S. and coalition forces out of Iraq (Hassan, 2008, pp. 278-279). Most of the opposition in Iraq came from ex-regime supporters and pro-Baathists, Sunni Iraqi Islamists, and foreign volunteer fighters (Hassan, 2008, p. 278). In particular, the Sunni Arab community began attacking the U.S.-backed Shia government and Shia civilians at mosques and holy sites (Zunes, 2006, p. 73). In response, the Shia majority Iraqi government and military responded to the insurgency with an intense counter-insurgency program and counter-terrorist operations aimed at the Sunni community. These operations led to a dramatic increase in the violence against the Sunni minority by the Shia majority. Human rights groups reported that hundreds of Sunni bodies showed signs of torture and summary execution every month (Zunes, 2006, ¶ 17). The perpetrators of these horrendous crimes were death squads under the control of the Ministry of the Interior. Further, the situation was exacerbated by the conduct of U.S. troops: their liberal use of force in heavily populated civilian areas and at check-points; torture; and, the detention of innocent young men because of suspicions of ties to the insurgency—which not only made the situation worse, it incited more people to join Sunni militant organizations (Zune, 2006, ¶ 18). One more result of this sectarian violence was a refugee crisis that has had widespread ramifications for not only Iraq, but for the entire region.

Amnesty International (AI, 2008), described how the Invasion of Iraq and the subsequent internal armed conflict had a disastrous effect on the lives of average Iraqis. Shia, Sunni, Christians, and others fled for their lives and now struggle to survive in host countries. As of 2008, it is estimated that 4.7 million Iraqis have been displaced, with 2.7 million people internally displaced in Iraq (Amnesty International [AI], 2008, pp. 2-3)—out of a total population estimated by the UN to be a little over 25 million. Each month it was estimated that there were 30,000 to 50,000 refugees fleeing Iraq and flowing into neighbouring countries (Goodman, 2007, ¶ 27). As journalist Nir Rosen explained:

> "Iraq has been changed irrevocably, I think. I don't think Iraq even— you can say it exists anymore. There has been a very effective, systematic ethnic cleansing of Sunnis from Baghdad, of Shias—from areas that are now mostly Shia. But the Sunnis especially have been a target...". (Goodman, 2007, ¶ 9)

Host countries are doing what they can to support these people, but the burden is too much for them. The international community has failed in both its moral and legal responsibility to aid these people. In fact, the response by the international community has been woefully inadequate at best, to downright apathetic at worst (AI, 2008, p. 2). Financial support to both host countries and aid organizations working to aid refugees is a pittance and not enough to forestall the growing crisis and UN agencies are in need of more funding to keep essential programs from floundering.

Figure 3.4

A photo contained in a Naval Criminal Investigative Service report shows a Marine inspecting a roadside scene near Haditha, Iraq, where five unarmed civilians were killed on 19 November 2005. Earlier that day, Marines stopped the white taxi in which the men had been riding, and then allegedly shot them after a bomb exploded nearby. The incident was the first on a day of violence in Haditha that left 24 civilians dead, among them women and children, and four Marines charged with murder. (Source: Wikimedia Commons, under a Creative Commons License.)

The lives these refugees have is hard as they are banned from working and humanitarian aid agencies are unable to keep up with the growing demand to support them. These people are faced with a dilemma: they can either stay in their host country in worsening conditions or return to Iraq where security issues and human rights violations are still of great concern (AI, 2008, p. 4). In 2007, there was an improvement in the security situation, but it is still neither safe nor suitable to return to Iraq. People are still being killed by armed militias, Iraqi Security Forces, multi-national military forces, and private military personnel (AI, 2008, p. 7).

However, what is truly startling is how the people fleeing also bring with them this sectarianism and their connection to militias in Iraq, to the host country (Politicstv, 2007, 2:50). This spread of sectarianism from Iraq to other countries in the region could de-stabilize the host countries as well. The conflict in Iraq has the potential to become more than simply an internal conflict: it could lead to a civil war that will not only destabilize Iraq, but the entire region (Politicstv, 2007, 3:30). Further, the current strategy of arming warlords and using them as security forces could tear the country in pieces in the future when U.S. and coalition eventually leave Iraq (Chomsky, 2009, ¶ 40). Along with this strategy, they have also created walled communities that are a refuge for displaced peoples. This strategy will only escalate tensions and accelerate the ethnic cleansing of the Sunni population. With all of the devastation that has been caused by the Iraq War, Iraq has essentially been changed to the point where it will never be the same.

Figure 3.5

A bullet can be seen being fired from the Apache helicopter while shelling the minivan that had come to aid the civilians shot by the Apache. (Source: from WikiLeaks' Collateral Murder video at http://www.collateralmurder.com. In the public domain.)

Figure 3.6

The children, Sayad and Doaha, were hit by shrapnel shells and fell unconscious due to their heavy wounds after a U.S. Apache helicopter rained fire down on their van, and fired a round through the windshield. (Source: Video still taken from WikiLeaks' Collateral Murder video as shown at http://www.collateralmurder.com. In the public domain.)

Conclusion

When the U.S. decided to invade Iraq, it had no idea what it was getting itself into and instead of stabilizing Iraq and promoting democracy, the U.S. has done the opposite and made the situation in Iraq far worse than it was before. The U.S. has not only destabilized Iraq, but possibly the entire region.

The over-reliance on coercive means to achieve goals can be counterproductive, as can clearly be seen in the case of Iraq. First, in the case of sanctions, instead of removing Saddam Hussein from power, the sanctions gave him more power and the subsequent invasion made an already volatile situation even more so. Iraq can be seen as the first case of a new model of preemptive war that can be used in the future against other "rogue" states that are deemed a threat to U.S. national security.

If Iraq is to be a model for preemptive war, then other states should be wary, because Iraq will be feeling the after affects of U.S. influence and war for decades to come. Also, it is a little ironic that the U.S. would be proponents of this new model of war, given the fact Japan was reviled for its preemptive attack at Pearl Harbour and was demonized for it. If the U.S. objective was truly to spread peace and democracy around the world, doing so at the barrel of a gun produces the opposite results. We must learn from our past and find a way to avoid these sorts of conflicts. If not, we will be doomed to repeating the past until our past ends us.

Notes

1   See the copy of "Protocol Additional to the Geneva Conventions of 12 August 1949, and relating to the Protection of Victims of International Armed Conflicts (Protocol I)" on the website of the United Nations High Commission for Refugees at:
   http://www.unhcr.org/refworld/docid/3ae6b36b4.html.

2   To see that portion of the interview with Albright, please see: http://www.youtube.com/watch?v=FbIX1CP9qr4. While some have said Albright's words were taken out of context, she in fact owned up to them and apologized, for example, on *Democracy Now!*—see:
   http://www.democracynow.org/2004/7/30/democracy_now_confronts_madeline_albright_on.

3   That cost has in fact been vast: see for example the website of the *Cost of War* project co-directed by anthropologist Catherine Lutz at Brown University, which finds that: "Conservatively estimated, the war bills already paid and obligated to be paid are $3.2 trillion in constant dollars. A more reasonable estimate puts the number at nearly $4 trillion". http://costsofwar.org/

References

Amnesty International (AI). (2008). *Rhetoric and Reality: The Iraq Refugee Crisis.* London: Amnesty International Publications.
   http://amnesty.org/en/library/asset/MDE14/011/2008/en/2e602733-42da-11dd-9452-091b75948109/mde140112008eng.pdf

BBC. (2008). Iraq War "Violated Rule of Law". *British Broadcasting Corporation,* November 18.
   http://news.bbc.co.uk/2/hi/uk_news/politics/7734712.stm

Bennett-Jones, O. (2004). "It was illegal," UN Secretary-General Kofi Annan on the Iraq war [Transcript].

http://news.bbc.co.uk/2/hi/middle_east/3661640.stm

Blix, H.. (2003). Hans Blix's Briefing to the Security Council. *The Guardian*, February 14.

http://www.guardian.co.uk/world/2003/feb/14/iraq.unitednations1

Burkeman, O., & Borger, J. (2003). War Critics Astonished as US Hawk Admits Invasion Was Illegal. *The Guardian*, November 20.

http://www.guardian.co.uk/uk/2003/nov/20/usa.iraq1

Chomsky, N. (2004). *Hegemony or Survival: America's Quest for Global Dominance*. New York: Henry Holt.

———. (2008a). Humanitarian Imperialism: The New Doctrine of Imperial Right. *ZNet*, September 15.

http://www.zcommunications.org/humanitarian-imperialism-by-noam-chomsky

———. (2008b). It's the Oil, Stupid! *Khaleej Times Online*, July 8.

http://www.khaleejtimes.com/DisplayArticleNew.asp?col=&section=opinion&xfile=data/opinion/2008/July/opinion_July32.xml

———. (2009). Modern Day American Imperialism: The Middle East and Beyond. *ZNet*, December 23.

http://www.zcommunications.org/modern-day-american-imperialism-the-middle-east-and-beyond-by-noam-chomsky

CNN. (2003). Bush Declares War. *CNN U.S.*, March 19.

http://articles.cnn.com/2003-03-19/us/sprj.irq.int.bush.transcript_1_coalition-forces-equipment-in-civilian-areas-iraqi-troops-and-equipment?_s=PM:US

Commisie van onderzoek besluitvorming Irak. (2010). Rapport Commissie Van Onderzoek Besluitvorming Irak.

http://openanthropology.files.wordpress.com/2010/06/rapport_commissie_i_267285a.pdf

Gellman, B. (1991). Allied Air War Struck Broadly In Iraq: Officials Acknowledge Strategy Went Beyond Purely Military Targets. *The Washington Post*, June 23.

http://www.envirosagainstwar.org/know/1991USHitCivilianTargets.pdf

Goodman, A. (2007). "Iraq Does Not Exist Anymore": Journalist Nir Rosen on How the U.S. Invasion of Iraq Has Led to Ethnic Cleansing, a Worsening Refugee Crisis and the Destabilization of the Middle East [Transcript]. *Democracy Now!* August 21.

http://www.democracynow.org/2007/8/21/iraq_does_not_exist_anymore_journalist

Halliday, D. (2000). UN Sanctions Against Iraq Only Serve U.S. Ambition. *The Irish Times*, August 11.

http://www.commondreams.org/views/081100-104.htm

Hassan, R. (2008). Global Rise of Suicide Terrorism: An Overview. *Asian Journal of Social Science*, 36, 271-291.

http://www.riazhassan.com/archive/Global_Rise_of_Suicide_Terrorism.pdf

Hayes, C. (1990). Confrontation in the Gulf; the Oilfield Lying Below the Iraq-Kuwait Dispute. *The New York Times*, September 3.

http://www.nytimes.com/1990/09/03/world/confrontation-in-the-gulf-the-oilfield-lying-below-the-iraq-kuwait-dispute.html?pagewanted=all&src=pm

Hirsch, A. (2010). Iraq Invasion Violated International Law, Dutch Inquiry Finds—Investigation into The Netherlands' Support for 2003 War Finds Military Action Was Not Justified Under UN Resolutions. *The Guardian*, January 12.
http://www.guardian.co.uk/world/2010/jan/12/iraq-invasion-violated-interational-law-dutch-inquiry-finds

International Commission of Jurists (ICJ). (2003). Iraq - ICJ Deplores Moves Toward a War of Aggression on Iraq. *International Commission of Jurists*, March 18.
http://web.archive.org/web/20090622163632/http://www.icj.org/news.php3?id_article=2770&lang=en

Kessler, G., & VandeHei, J. (2004). Misleading Assertions Cover Iraq War and Voting Records. *The Washington Post*, October 6, A15.
http://www.washingtonpost.com/wp-dyn/articles/A10244-2004Oct5.html

Lock, E. (2010.) Soft Power and Strategy: Developing a 'Strategic' Concept of Power. In I. Parmar & M. Cox (Eds.), *Soft Power and US Foreign Policy: Theoretical, Historical and Contemporary Perspectives* (pp. 32-50). New York: Routledge.

Lynch, J. (1997). Iraq Sanctions. *TED Case Studies*, 7(1), January.
http://www1.american.edu/projects/mandala/TED/iraqsanc.htm

Politicstv. (2007). New America's Nir Rosen on Iraqi Refugee Crisis [Video]. May 14.
http://www.youtube.com/watch?v=-xcCOOEUCkY

Nagy, T. (2001). The Secret Behind the Sanctions: How the U.S. Intentionally Destroyed Iraq's Water Supply. *The Progressive*, September.
http://www.progressive.org/mag/nagy0901.html

Powell, C. (2003a). Full text of Colin Powell's Speech—U.S. Secretary of State's Address to the United Nations Security Council (Part 1). *The Guardian*, February 5.
http://www.guardian.co.uk/world/2003/feb/05/iraq.usa

————. (2003b). Full text of Colin Powell's Speech—U.S. Secretary of State's Address to the United Nations Security Council (Part 2). *The Guardian*, February 5.
http://www.guardian.co.uk/world/2003/feb/05/iraq.usa3

————. (2003c). Full text of Colin Powell's Speech—U.S. Secretary of State's Address to the United Nations Security Council (Part 3). *The Guardian*, February 5.
http://www.guardian.co.uk/world/2003/feb/05/iraq.usa2

Reuters. (1990). The Iraqi Invasion; In Two Arab Capitals, Gunfire and Fear, Victory and Cheers. *The New York Times*, August 3.
http://www.nytimes.com/1990/08/03/world/the-iraqi-invasion-in-two-arab-capitals-gunfire-and-fear-victory-and-cheers.html?pagewanted=all&src=pm

Rose, C. (2004). A Conversation with Paul Volcker about the United Nations' Oil-for-Food program for Iraq [Video]. *Charlie Rose*, November 24.
http://www.charlierose.com/view/interview/1168

Russert, T. (2003). Sunday, September 14, 2003—Guest: Dick Cheney, vice president [Transcript]. *Meet the Press*.
http://msnbc.msn.com/id/3080244/

Smith, M. (2010). Revealed: Jack Straw's Secret Warning to Tony Blair on Iraq. *The Sunday Times*, January 17.
http://www.timesonline.co.uk/tol/news/politics/article6991087.ece

United Nations (UN). (1991). Resolution 687, United Nations Security Council, April 3.
http://www.fas.org/news/un/iraq/sres/sres0687.htm

Wilmhurst, E. (2005). Wilmhurst Resignation Letter. *BBC*, March 24.
http://news.bbc.co.uk/2/hi/uk_news/politics/4377605.stm

Zakaria, F. (2008). *The Post-American World*. New York: W.W. Norton

Zunes, S. (2006). The U.S. Role in Iraq's Sectarian Violence. *Antiwar.com*, March 7.
http://www.antiwar.com/orig/zunes.php?articleid=8668

————— . (2007). The United States in Iraq: The Consequences of Occupation. *International Journal of Contemporary Iraqi Studies*, 1(1), 57-75.

CHAPTER FOUR

TORTURE AND THE GLOBAL WAR ON TERROR

Natalie Jansezian

T he primary intent of this chapter is to discuss and analyze the torture tactics executed by American soldiers in the global war on terror. This chapter will commence with a brief introduction regarding the global war on terror as declared by the administration of President George W. Bush, as well as its main objectives. In addition, I will briefly discuss the scandal of Abu Ghraib that emerged regarding the acts of torture inflicted upon Iraqi detainees by U.S. troops. More specifically, I will focus the discussion on the issue of torture; what it is, the forms of torture that have been used and how torture has been used against detainees. This chapter will also take a look into the authorization of the Bush administration regarding the use of torture and abuse against detainees. In other words, although multiple reports as well as photographs indicate that the U.S. did, in fact, use torture in an attempt to extract information from detainees, the Bush administration knowingly allowed and authorized the use of torture during the process of interrogation. This discrepancy is one significant point that I will raise in this chapter. In order to do so, I will discuss and elaborate on the Geneva Convention Relative to the Treatment of Prisoners of War and the Convention against Torture and Other Cruel, Inhuman, or Degrading Treatment or Punishment.

Furthermore, another point that I will raise is the notion of the terrorist and what constitutes a terrorist. Although there is a conjured image of what a stereotypical terrorist is, a terrorist may take different forms. In this paper I will elaborate on how the vengeful actions and the disregard for the Convention against Torture and Other Cruel, Inhuman or Degrading Treatment or Punishment taken on part of the U.S. are acts of terrorism. In other words, the U.S. military and the Bush administration among the other groups of individuals who approved and imposed the malicious and immoral acts on detainees are terrorists.

The objective of this paper is two-fold: first, I will raise the issue of the discrepancy that exists between what the Bush administration said and the actual events that took place as well as the images that have emerged concerning the illegal use of torture. Second, I will prove that the American individuals who disregarded the Conventions and continued to execute acts of torture on detainees are terrorists.

The Global War on Terror

Close to 3,000 people were killed in the terrorist attacks on 11 September 2001. The Bush administration declared its global war on terror soon after these atrocious attacks occurred as a means to fight the organization that perpetrated these attacks (Record, 2003, p. 1). According to Record (2003, p. v), the objectives of the global war on terror include the "destruction of al-Qaeda and other transnational terrorist organizations, the transformation of Iraq into a prosperous, stable democracy, the democratization of the rest of the autocratic Middle East" and the "eradication of terrorism as a means of irregular warfare". However, Record (2003, p. v) expresses that the objectives of the global war on terror are "unrealistic and condemn the United States to a hopeless quest for absolute security". Moreover, the objectives of the global war on terror are "politically, fiscally, and militarily unsustainable" (Record, 2003, p v).

"War" has seemingly been embraced in American political discourse "as a metaphor for dealing with all kinds of enemies, domestic and foreign" (Record, 2003, p. 2). In fact, Record (2003) argues that *war* is "perhaps the most over-used metaphor in America" (2003, p. 2). This suggests that at least since the 9/11 attacks and the declaration of the so-called global war on terror, Americans have become more immune to the implications of the widely used notions of war and terror, and the meaning of torture. Moreover, the term "war," used as it once was to refer to conflict between states—wars that had clear beginnings and endings—clearly loses its conceptual clarity when used to discuss dispersed, non-state actors, and a permanent mode of conflict (Record, 2003, pp. 3, 4).

The Abu Ghraib Scandal

In response to the terrorist attacks on 9/11, Bush ordered interrogators to use "enhanced interrogation techniques," in order to extract vital information from the Iraqi detainees (CNN, 2009, ¶ 9). However, these "enhanced interrogation techniques" included illegal acts of torture. The torture that was used during the interrogation process occurred in the prison in Abu Ghraib, which rapidly became a controversy.

Abu Ghraib, west of Baghdad, is a prison well known for being a place where torture and executions were common under Iraqi authorities (*The New York Times* [NYT], 2011, ¶ 1). In the year 2004, images surfaced consisting of American soldiers physically and sexually abusing and torturing Iraqi prisoners of war which ultimately created an uproar among segments of the American public and the greater population across the globe (NYT, 2011, ¶ 2). Consequently, after the emergence of these explicit photos, President George W. Bush proposed a thorough investigation to take place in order to uncover the true events that occurred in the Abu Ghraib prison (NYT, 2011, ¶ 4), thus creating an illusion of unawareness at the top levels of the administration. Ultimately, only nine American soldiers were found guilty for the acts of torture and abuse they imposed on Iraqi prisoners (NYT, 2011, ¶ 5).

Although the acts of torture and abuse that were inflicted on Iraqi detainees by American soldiers were revealed, the true nature of the occurrences that took place would not be known to the public until WikiLeaks (2010) uncovered the secret military documents which possessed vital information regarding allegations made by the prisoners of war and an astounding misrepresentation of the actual death total of Iraqi civilians.

Torture

According to Article 1 of the Convention against Torture and Other Cruel, Inhuman or Degrading Treatment or Punishment established by the United Nations,

> "torture means any act by which severe pain or suffering, whether physical or mental, is intentionally inflicted on a person for such purposes as obtaining from him or a third person information or a confession, punishing him for an act he or a third person has committed or is suspected of having committed, or intimidating or coercing him or a third person, or for any reason based on discrimination of any kind, when such pain or suffering is inflicted by

or at the instigation of or with the consent or acquiescence of a public official or other person acting in an official capacity". (UN, 1984)

The methods used against detainees by American interrogators in order to obtain information or a confession are identified as torture.

The allegations of torture and abuse imposed on civilian and other detainees by U.S. officials and troops surfaced through the internet. In fact, "WikiLeaks has published almost 400,000 U.S. military logs, mainly written by soldiers on the ground, detailing daily carnage in Iraq since the 2003 U.S.-led invasion: detainees abused by Iraqi forces, insurgent bombings, sectarian executions and civilians shot at checkpoints by U.S. troops" (Associated Press [AP], 2010, ¶ 3). These reports contained in-depth details and accounts of U.S. troops imposing abuse and torture on foreign detainees in order to extract any information which may be of interest and possess importance for the American officials as well as the greater population. The contents of these documents will be further discussed and explored later on.

President Bush signed a series of directives authorizing the CIA to conduct a covert war against Al Qaeda, directives that "empowered the agency to kill or capture Al Qaeda leaders" (NYT, 2010, ¶ 1). Attorney General Eric Holder requested for federal prosecutor John Durham to investigate whether the interrogations of suspected terrorists conducted by the CIA were illegal (CNN, 2009, ¶ 1). After extensively viewing CIA interrogation tapes, Durham found that "U.S. interrogators threatened a captured al Qaeda operative with a power drill to try to scare him into giving up information" (CNN, 2009, ¶ 27). In addition, Durham reported that U.S. interrogators threatened to kill the detainee's children in an attempt to retrieve information from the captive's (CNN, 2009, ¶ 27). Durham's report also stated that "unauthorized detainee interrogation methods also included, among other things, 'the making of threats, blowing cigar smoke, employing certain stress positions, the use of a stiff brush on a detainee, and stepping on a detainee's ankle shackles'" (CNN, 2009, ¶ 31).

Moreover, mock executions were staged, which are illegal according to the U.N. Convention against Torture, in order to gain information from the captives (CNN, 2009, ¶ 32). For instance, when the detainees were escorted to another room they would pass by a guard who was lying lifelessly on the ground covered with a hood over his head as if he had been shot (CNN, 2009, ¶ 33). This made it seem as if detainees were being executed if they did not provide any useful information to the interrogators.

Figure 4.1

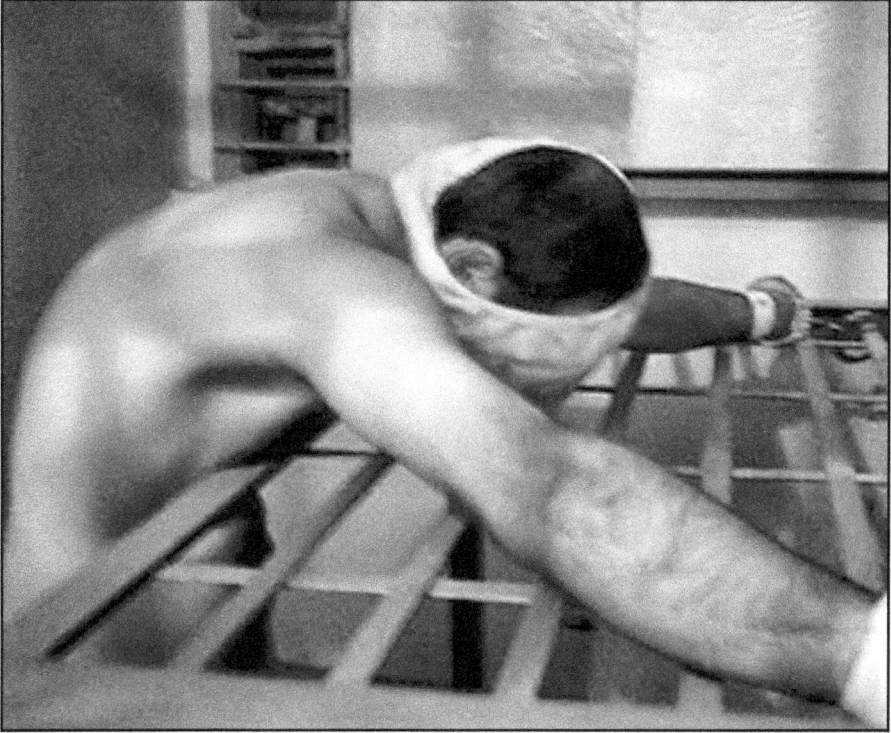

Detainee is handcuffed nude to a bed and has a pair of panties covering his face. Photograph is taken from the entrance of the cell. 1:53am, 20 October 2003, Abu Ghraib prison, Iraq. (Source: provided by Wikimedia Commons under a Creative Commons License, as it is ineligible for copyright.)

Figure 4.2

Blood on the floor of a cell in Abu Ghraib, Iraq, the extensive smear showing that a bloodied captive had been dragged from his cell. (Source: provided by Wikimedia Commons under a Creative Commons License, as it is ineligible for copyright.)

Figure 4.3

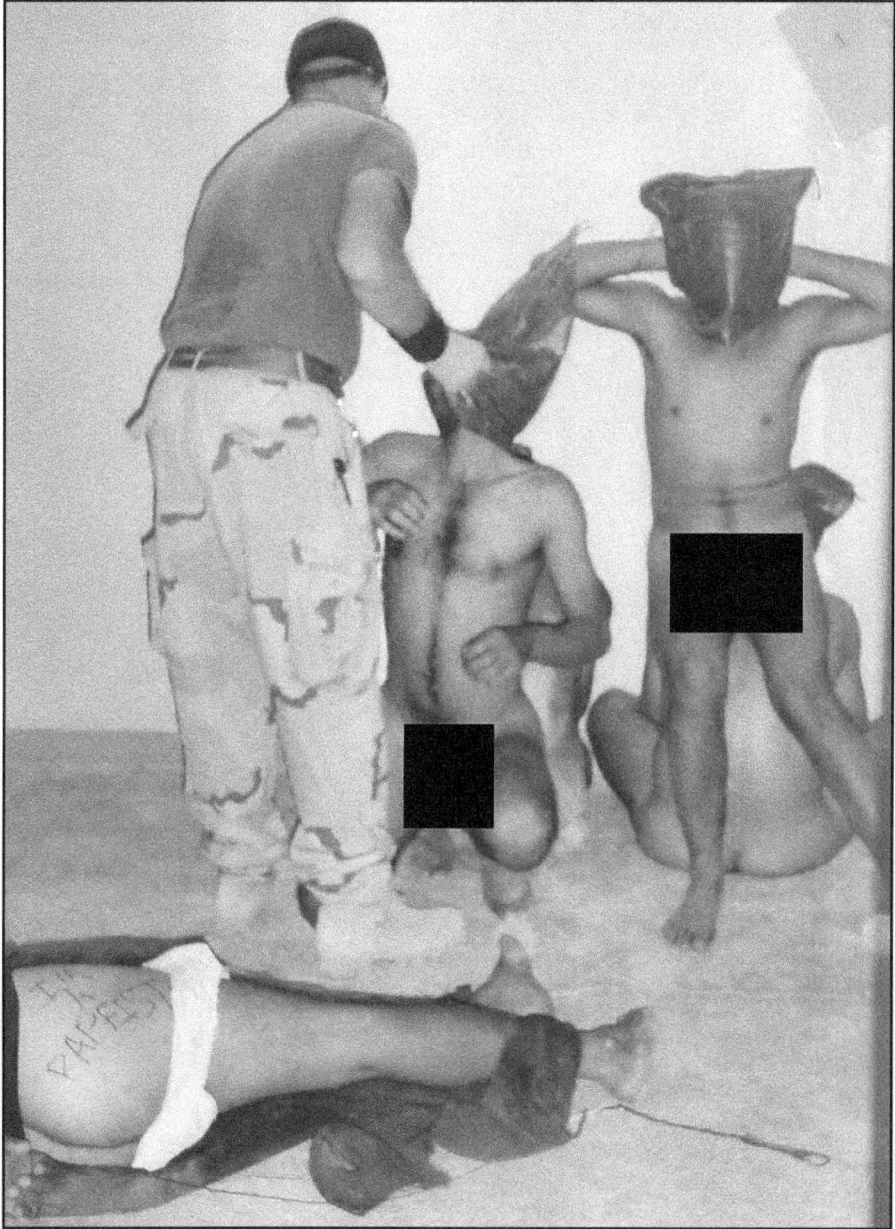

Detainees being hooded and humiliated in Abu Ghraib prison, Iraq. One has "I'm a rapeist [sic]" written on his backside. (Source: provided by Wikimedia Commons under a Creative Commons License, as it is ineligible for copyright.)

According to CNN (2009, ¶ 37), "the interrogations took place in the CIA's secret prisons before 2006, when Bush moved all detainees from such facilities to the detention center in the U.S. Navy base in Guantanamo Bay, Cuba". Lowry (2011, ¶ 1), explains that approximately 200 prisoners were detained at the detention center in Guantanamo Bay, the facility that Amnesty International calls "a global symbol for injustice and abuse".

Figure 4.4

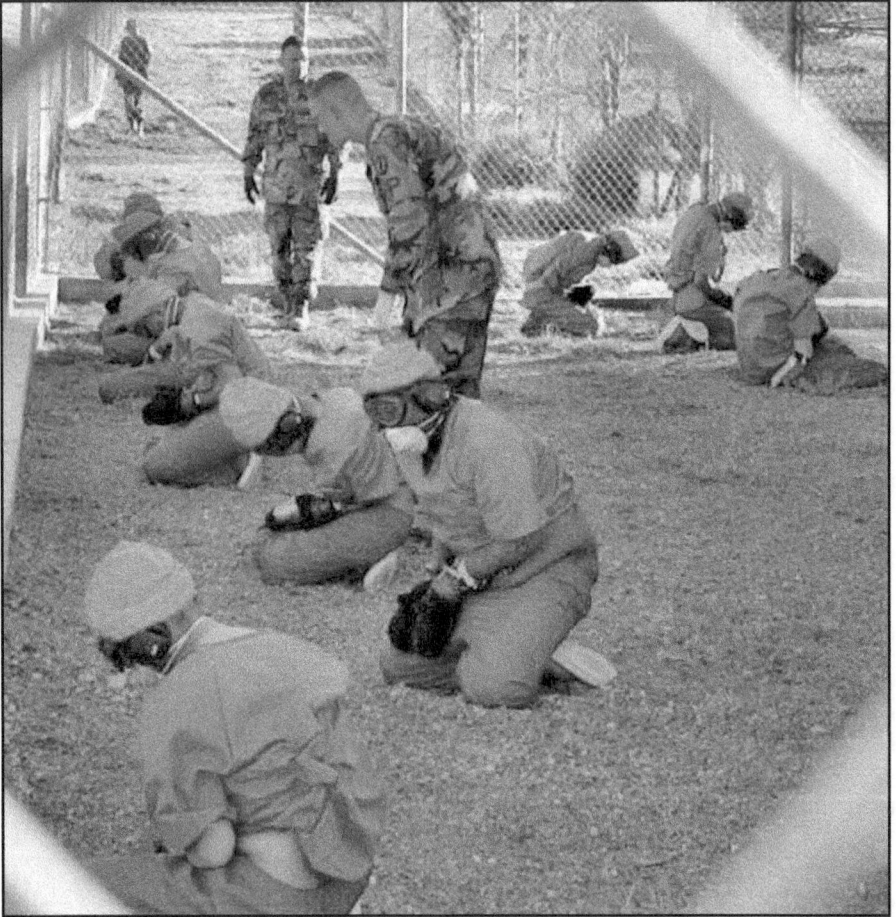

Detainees in orange jumpsuits sit in a holding area under the watchful eyes of Military Police at Camp X-Ray at Naval Base Guantanamo Bay, Cuba, during in-processing to the detention camp on 11 January 2002. (Source: U.S. Department of Defense photo by Petty Officer 1st class Shane T. McCoy, U.S. Navy. Provided by Wikimedia Commons under a Creative Commons License.)

It is clear that the CIA interrogators used immoral and callous methods in order to interrogate and extract information from the detainees. The torture techniques that were used against detainees were borrowed from an "American military training program modeled on the torture repertories of the Soviet Union and other cold-war adversaries" (NYT, 2010, ¶ 3). This training program known as SERE (Survival, Evasion, Resistance and Escape) included techniques such as water boarding which has been a "historically well-known torture" (Nance, 2009, ¶ 5). Water boarding consisted of interrogators pouring water "over a detainee's nose and mouth for a total of 12 minutes a day;" also, "interrogators were instructed to start pouring water right after a detainee exhaled, to ensure he inhaled water, not air, in his next breath" (Benjamin, 2010, ¶ 3). Another torture tactic that was borrowed from the SERE training program and was used by interrogators was sleep deprivation. According to Mayer, detainees were deprived of sleep on a continuous basis, often by being bombarded by loud, unbearable sounds and bright, nauseating light (Mayer, 2008, p. 169). According to SERE personnel, it was more difficult for detainees to endure loud, continuous noises, specifically of babies crying, than it was to experience water boarding (Mayer, 2008, p. 170). Finally, the third most popular technique employed by interrogators is what is known as extraordinary rendition. This process involves a detainee being sent to another country where the use of torture is legal (Pither, 2008, p. 101).

Not only did American troops and interrogators fail to abide by the Convention against Torture and Other Cruel, Inhuman or Degrading Treatment or Punishment, but they also neglected the rights of the prisoners of war that have been established by the Geneva Convention Relative to the Treatment of Prisoners of War. According to Article 13 of this Convention, as taught to U.S. soldiers in Training Circular 27-10-2, prisoners of war "must at all times be protected, particularly against acts of violence or intimidation, and against insults and public curiosity" (Department of the Army [DoA], 1991, p. 5). In addition, the Convention explicitly states that the detaining power must provide the prisoner with food, clothing, medical aid and safe drinking water (DoA, 1991, p. 6). In another point, "under the Geneva Convention, the detaining power cannot subject a prisoner to *physical or mental torture, or any other force*, to secure information [emphasis added]" (DoA, 1991, p. 13). Evidently, U.S.-appointed interrogators ignored these two Conventions.

In addition to the controversial scandal of Abu Ghraib, other allegations of abuse were made against U.S. troops by Iraqi detainees. In fact, the U.S. made a commitment to eradicate and prevent the use of torture on civilians and prisoners from occurring again; however, according to the Bureau of Investigative Journalism (BIJ), which primarily based its reports on the WikiLeaks Iraq War Logs, 303 allegations of torture and abuse employed on Iraqi prisoners by United States forces were reported after 2004 (BIJ, 2010d, ¶ 1). Out of these reports, 42 were allegations that involved malicious forms of abuse that were borrowed from the SERE training program, such as beatings, water boarding, electric shocks and mock executions (BIJ, 2010d, ¶ 2).

For example, in an account by a detainee in 2005, he claimed that "he was blindfolded and beaten after being arrested and detained by U.S. Marines at a checkpoint" (BIJ, 2010d, ¶ 5). In another claim, a detainee stated that U.S. Marines were videotaping the beatings and torture throughout the duration of the abuse:

"Redacted Report 13-5: 12 November 2006

Two marines (Corporal and Lance Corporal) allegedly videotaped themselves...holding a knife to a detainee's throat and a M9 to the detainees head".[1]

Medical evidence verified most of the claims of abuse, in 20 of the 30 cases where medical investigations took place (BIJ, 2010d, ¶8):

"Redacted Report 13-2: 11 July 2006

The detainee stated that after he was flex-cuffed, one person sat on his chest and another on his legs...the person punched him in the back of the head, picked up his head and slapped him, and put a plastic pipe in his mouth...The persons conducting the questioning also kicked him on the sides of his body...after this the persons put a bag over his head....The medic concluded that the detainee did have injuries to his back that were consistent with abuse. A statement was taken from the detainee. Pictures were taken to document the abuse".[2]

Not all of the reports accuse the U.S. troops of using physical violence; rather, other reports state that U.S. soldiers inflicted mental abuse on the detainees as well. For example, the U.S. troops tactically instilled "fear of death" in the prisoners throughout the interrogation process (BIJ, 2010d, ¶ 10). This refers to the notion of the U.S. troops threatening to turn over the detainees to the Iraqi authorities, who, consequently, were known to engage in torture as a matter of routine, as well as acting upon these threats. In doing so, U.S. troops clearly violated the Geneva Convention. These allegations go strictly against the promise that the U.S. government made in regards to the elimination of torture during the interrogation process and the assurance that these reported allegations would be thoroughly investigated and improved on. It is evident that allegations were being reported; however, what is not known is what actions were taken upon after the reports were being gathered (BIJ, 2010d, ¶ 13). According to the Bureau of Investigative Journalism (2010d, ¶ 13), "the Department of Defense has refused to comment on specific cases".

Not only were U.S. officials and interrogators discovered to be using ruthless tactics to extract information from prisoners, but security services of nations such as Iraq were also discovered to be using methods of torture on their own citizens. It became known that U.S. troops were aware of the abuse and torture that was being executed by the Iraqi security services on Iraqi civilians. However, the U.S. troops failed to exercise their power and their responsibility to protect; in fact, U.S. troops were ordered not to conduct an investigation concerning the torture imposed by Iraqi forces: two orders, FRAGO 242 and FRAGO 039 specifically ordered either no further investigation or no further action (BIJ, 2010c). In addition, secret military files show that "U.S. soldiers witnessed, or were told of, more than 1,300 cases of detainee abuse by Iraqi authorities"; however, "following the Abu Ghraib scandal of 2004, they were given explicit orders not to investigate unless coalition personnel were involved" (BIJ, 2010c, ¶ 2). This further illustrates the point that U.S. forces were aware, through observations and dialogue, that Iraqi prisoners were being violently tormented and abused by Iraqi forces and with this knowledge, U.S. forces did not attempt to aid or prevent these acts of torture.

The documents released by WikiLeaks explain that "U.S. authorities failed to investigate hundreds of reports of abuse, torture, rape and even murder by Iraqi authorities" (Newscore, 2010, ¶ 1). Indeed, while "six of these reports apparently end with the detainee dying because of the alleged torture," in spite of that fact, "the claims were never properly investigated" (Newscore, 2010, ¶ 6).

According to the Bureau of Investigative Journalism (2010a, ¶ 1), "U.S. forces discovered 173 incarcerated men in a secret detention centre run by the Iraqi Ministry of Interior. Many showed signs of brutal torture". On 13 November 2005, an Iraq War log indicated that approximately 95 detainees were being held in a room and "were sitting cross-legged with blindfolds, all facing the same direction. According to one of the detainees questioned on site, 12 detainees have died of disease in recent weeks" (BIJ, 2010a, ¶ 3). Another document explains that a detainee's "hands were bound/shackled and he was suspended from the ceiling; the use of blunt objects (pipes) to beat him on the back and legs; and the use of electric drills to bore holes in his legs" (BIJ, 2010b, ¶ 2). Another detainee claimed that,

> "his interrogators kicked, punched, slapped him on the face, stomach, legs, and also electrocuted his hands, left ear, and genital area for approximately 15-30 minutes....[the] detainee said that his mouth was covered...because the Americans might hear his screams. Detainee was then taken back to his cell...he fainted and was carried to his bed." (BIJ, 2010b, ¶ 4)

In another account, a detainee reportedly received third degree chemical burns and skin decay from when an Iraqi army lieutenant colonel "tortured him by pouring chemicals on his hands, cut his fingers off, and hid him when Coalition Forces visited" (BIJ, 2010b, ¶ 8). These are only a few reported accounts out of 1,365 cases "categorized in the military records as potential detainee abuse by Iraqi authorities" (BIJ, 2010b, ¶ 3).

Immediately after the discovery of the incarcerated men and the mistreatment they endured, an investigation and inspection was summoned by the U.S. forces. Among its findings, "the U.S.-Iraqi inspection claimed it found no torture in the seven facilities investigated. But while these inspections were going on American troops were reporting case after case of alleged abuse and torture by Iraqi personnel elsewhere" (BIJ, 2010a, ¶ 4). In addition to this point, "a coalition spokesman told reporters: 'The facilities were, by our standards, overcrowded, but the people being held at those facilities were being properly taken care of; they were being fed, they had water, they were taken care of. So no abuse, no evidence of torture in those facilities'" (BIJ, 2010c, ¶ 16). However, the U.S military logs that were distributed by WikiLeaks contradicted the findings of the U.S.-Iraqi inspection. According to the WikiLeaks files, "during the same four-month period of the inspections, the Bureau of Investigative Journalism has identified 76 allegations of abuse by Iraqi security forces that were all reported by U.S. forces. While the cases do not relate to the specific sites of the coalition inspections, they prove the U.S. was aware of a prevailing use of torture by Iraqi security forces even when the results were being compiled" (BIJ, 2010a, ¶ 8). In other words, U.S. forces were well aware of the torture and abuse that Iraqi civilians were enduring at the hands of the Iraqi forces. In addition, U.S. forces deliberately did not swiftly respond or intervene in the mistreatment of Iraqi prisoners.

State of Exception, State of Denial, State of Continuity

Given secret military documents that have been leaked and published by WikiLeaks and the serious allegations that have been made against U.S. troops regarding the use of ruthless and malicious acts of terror and torture, it is clear that U.S. officials would have a hard time denying these accusations and hard evidence. CNN (2009, ¶ 9) reports that in the years following the 9/11 terror attacks, President George W. Bush gave his consent and approval concerning "enhanced interrogation techniques" for terror suspects. These "enhanced interrogation techniques" included methods such as water boarding, physical and mental abuse, sleep deprivation and extraordinary rendition among others. However, several years after the 9/11 terror attacks and after the Abu Ghraib scandal, President George W. Bush gave a speech on the United Nations International Day in Support of Victims of Torture.

In his speech, Bush reaffirmed the commitment that the United States made as a nation to the "worldwide elimination of torture" (Bush, 2004, ¶ 1). Bush continued to state that the "non-negotiable demands of human dignity must be protected without reference to race, gender, creed, or nationality. Freedom from torture is an alienable human right, and we are committed to building a world where human rights are respected and protected by the rule of law" (2004, ¶ 1). Bush reminded the American population of how the United States had joined forces with 135 other nations in order to authorize the Convention against Torture and Other Cruel, Inhuman or Degrading Treatment or Punishment (2004, ¶ 2). In addition, Bush stated that the "United States also remains steadfastly committed to upholding the Geneva Conventions" (2004, ¶ 2). In this presidential statement, Bush promised that the U.S. "will investigate and prosecute all acts of torture and undertake to prevent other cruel and unusual punishment in all territory under our jurisdiction" (2004, ¶ 2).

Bush's speech took place right after the revelation of photographs from the Abu Ghraib scandal. Bush recognized and acknowledged the fact that actions of torture and terror were imposed on Iraqi civilians by U.S. troops. Bush stated:

> "The American people were horrified by the abuse of detainees at Abu Ghraib prison in Iraq. These acts were wrong. They were inconsistent with our policies and our values as a Nation. I have directed a full accounting for the abuse of the Abu Ghraib detainees, and investigations are underway to review detention operations in Iraq and elsewhere." (2004, ¶ 4)

Figure 4.5

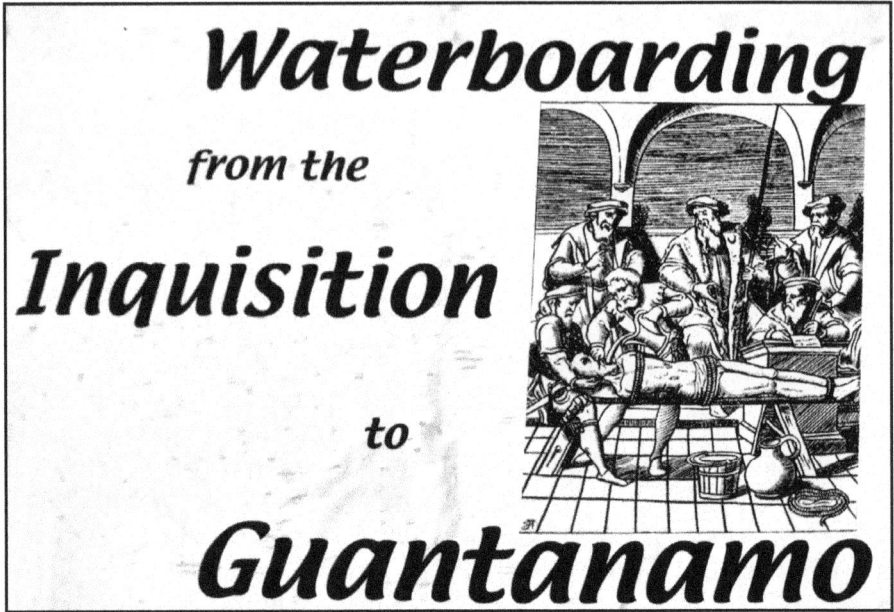

"Waterboarding, from the Inquisition to Guantanamo." A poster used for the Torture Abolition and Survivors Support Coalition and Amnesty International march in Washington, DC, 27 March 2007. (Source: provided by Wikimedia Commons under a Creative Commons License.)

Bush then defended the American troops by explaining that,

> "it is often American men and women in uniform who fight for the freedom of others from tyrannical regimes that routinely use torture to oppress their citizens. From Nazi Germany to Bosnia, and Afghanistan to Iraq, American service members have fought to remove brutal leaders who torture and massacre. It is the American people and their contributions that have helped to rebuild these traumatized nations to give former victims hope." (2004, ¶ 5)

What Bush failed to recognize and mention is that American troops have disrespected and disobeyed the Conventions they agreed to which were meant to protect and defend human rights. Rather, U.S. troops imposed torture and death threats on innocent Iraqi civilians. Moreover, the disregard for the Conventions on the part of the U.S. troops is due to the reason that Bush ordered them to use "enhanced interrogation techniques".

Bush concluded his speech by stating that,

"the United States will continue to take seriously the need to question terrorists who have information that can save lives. But we will not compromise the rule of law or the values and principles that make us strong. Torture is wrong no matter where it occurs and the United States will continue to lead the fight to eliminate it everywhere". (Bush, 2004, ¶ 8)

Bush gives the impression that he was astounded by the discovery that U.S. troops were using torture as an interrogation technique even though he authorized the use of "enhanced interrogation techniques" himself. In fact, Bush's speech completely denies his participation and authorization of torture, making it seem as if it were an isolated incident, an aberration, rather than a matter of policy.

Not only were American officials violating the Conventions that they were supposedly committed to, but they were also providing misleading information regarding the death total in Iraq. The Iraq War documents leaked to WikiLeaks contain 15,000 civilian deaths that have not been previously reported; Iraq Body Count (IBC) calculated from using the documents that over 150,000 violent deaths have been recorded since March 2003, with more than 122,000 (80%) of them civilian (IBC, 2010a, ¶ 1; 2010b, ¶ 3). In the words of one journalist (Fermino, 2010, ¶ 5), "the secret documents contain a litany of damaging info to the U.S. military, including reports of widespread detainee abuse by Iraqis and a higher than previously believed number of civilian deaths". In addition, the documents expressed that although the physical and mental abuse imposed by U.S. troops was significant, the torture imposed by Iraqi officials was much worse; however, U.S troops failed to intervene and ignored their suspicions concerning the mistreatment of civilians by Iraqi security forces (Fermino, 2010, ¶ 7).

One of the most damning assessments of U.S. human rights violations in its self-declared "war on terror," came from a high-level international panel of human rights and legal experts organized by the International Commission of Jurists, whose findings were published in a report by the Eminent Jurists Panel on Terrorism, Counter-terrorism and Human Rights (ICJ, 2009). One striking section in particular was widely quoted in the media (see for example AP, 2009):

"the United States, one of the world's leading democracies, has adopted measures to counter terrorism that are inconsistent with established principles of international humanitarian law and human rights law. Erroneously conflating acts of terrorism with acts of war, the United States Government proclaimed a 'war on terror', thereby misapplying war rules to situations not entailing armed conflict as understood by international humanitarian law. In genuine settings of warfare, it distorts, selectively applies and ignores otherwise binding rules, including relevant principles of human rights law. Other States have been complicit in some of the practices that have flowed from the war paradigm, and it is vital that the serious human rights violations that have occurred now be repudiated and remedied. The damage done to the rule of law must be repaired and the importance and value of upholding international humanitarian law and human rights law during all armed conflicts must be re-affirmed". (ICJ, 2009, p. 160)

The U.S. and other countries that used torture in the so-called war on terror "seriously damaged respect for basic human rights," said the ICJ's panel (AP, 2009, ¶ 1). At the very least we can conclude that Bush's 2004 speech was entirely misleading and disingenuous.

The American Terrorist

The acts of torture described thus far were committed, as we know, in the name of fighting "terrorism". It is important that we get a handle on the meaning of the concept of terrorism. Here I will focus on the various definitions of terrorism and what constitutes a terrorist as provided by sources such as the U.S. Department of Defense itself. When analyzing these several definitions of terrorism it is clearly evident that the U.S. did engage in terrorism, and we can analyze and relate the definitions of terrorism to the suffering of detainees at the hands of U.S. troops. I thus examine how U.S. officials granted and gave consent for U.S. soldiers to act in a threatening and violent manner, which thus constitutes "terrorism" in the U.S. government's own terms.

According to Record (2003, p. 6), in a study conducted in the year 1988, 109 definitions of terrorism were found (6). Record concluded that the only attributes that were consistent within the 109 definitions were that "terrorism involves violence and the threat of violence" (2003, p. 6). Furthermore, "the current U.S. national security strategy defines terrorism as simply 'premeditated, politically motivated violence against innocents'" (Record, 2003, p. 6). According to the U.S. Department of Defense, terrorism is defined as the "calculated use of unlawful violence to inculcate fear; intended to coerce or intimidate governments or societies in pursuit of goals that are generally political, religious, or ideological" (Record, 2003, p. 7). The U.S. National Strategy for Combating Terrorism places similar emphasis on terrorism "as a non-state phenomenon directed against the state and society; terrorism is 'premeditated, politically motivated violence perpetrated against non-combatant targets by subnational groups or clandestine agents'" (Record, 2003, p. 7). However, Record (2003, p. 7) points out that these definitions of terrorism exclude "state terrorism", which is a problem in suggesting that terrorism is inherently different when performed by subnational groups. Yet, it is clearly evident that the U.S. troops, U.S. officials, the Bush administration along with any other group of individuals who authorized or engaged in torture of detainees, including innocent Iraqi civilians for the most part, are terrorists.

Moreover, the actions of the U.S. troops and the authorization by U.S. officials to use torture on Iraqi civilians completely conform to the U.S. Department of Defense's indication of what defines an act of terrorism. As previously mentioned, the Defense Department declares terrorism as an "unlawful" act of violence in order to "inculcate fear" (Record, 2003p. 7). Clearly, the torture tactics that were used such as water boarding, mock executions, threats, physical, mental and sexual abuse, sleep deprivation and extraordinary rendition among others, were unlawful acts that defied the Conventions that the U.S had made and to which it declared itself to be fully and wholeheartedly committed. Furthermore, the torture tactics that were used were purposively imposed on the detainees in order to inculcate fear. In doing so, the U.S. troops were in "pursuit of goals that are generally political, religious or ideological" (Record, 2003, p. 7). In this case, the objective of U.S. troops was to instill fear in the detainees in order to get them to reveal relevant and vital information.

While U.S. troops, with the authorization from U.S. officials and the Bush administration, performed acts of terror, the Western mainstream tended not to portray the U.S. forces as terrorists; rather, they are significantly glorified and labeled as "heroes" in many instances. This is due to the reason that "the administration has cast terrorism and terrorists as always the evilest of evils, what the terrorist does is always wrong and what the counter-terrorist has to do is defeat them is therefore invariably, necessarily right" (Record, 2003, p. 7)—and the U.S. claims to be performing in the role of "counter-terrorist" even when it is the one initiating the terror against innocent civilian detainees. Of course this can be turned the other way too: those fighting against the U.S. believe the Americans to be the "evilest of evils" and they view themselves as doing the right thing by protecting their land and citizens from these intruders. As Record (2003, p. 9) stated, "one man's terrorist can in fact be another's patriot".

Conclusion

Ever since the attacks that occurred on 11 September 2001, the U.S. has been incessantly fighting in an on-going battle dubbed the "global war on terror". In order to retrieve vital information regarding the reasons and objectives of the terrorist attacks on America, as well as additional information, "President George W. Bush authorized 'enhanced interrogation techniques' for terror suspects" (CNN, 2009, ¶ 9). As a result, the torturous events that occurred in the Abu Ghraib prison were revealed and the American population, along with the rest of the world, was made aware of the immoral and malicious acts that the U.S. troops were ordered to commit. After the public was made aware of the Abu Ghraib controversial scandal, President Bush released a statement that expressed his dismay and disbelief to the torture tactics that were employed on Iraqi detainees. According to Bush, he was, apparently, unaware of the illegal interrogating practices that were taking place within the confines of the Abu Ghraib prison. Consequently, Bush reminded the American public that investigations would commence immediately and that the U.S. would remain committed to the promise they made to the Convention against Torture and Other Cruel, Inhuman or Degrading Treatment or Punishment, as well as the Geneva Conventions. However, once WikiLeaks revealed secret military documents to the population it became obvious that the U.S. had knowingly and willingly disobeyed the Conventions. Moreover, the evidence shows that not only did the U.S. troops and members of the CIA commit malicious acts of torture on Iraqi detainees because they were authorized to do so by the Bush administration, but the U.S. troops also neglected and ignored the brutal attacks that were being inflicted on Iraqi prisoners by the Iraqi forces because they were ordered not to investigate. Due to the evidence that has been provided through the exposure of the secret U.S. military documents as well as the numerous allegations made by detainees and the physical and medical proof that came along with it, it is evident that the U.S. failed to abide by the Conventions to which it, supposedly, was committed. This was not just a failure to act, as in negligence, but a willful transgression against the very norms which the U.S. swore to uphold.

When we examine the most basic, and least self-serving definitions of terrorism that have been produced by U.S. government agencies themselves, it is evident that the U.S. perfectly falls under the category of committing acts of terrorism against other nations and innocent civilians.

Notes

1 See http://www.iraqwarlogs.com/PDF/13/5.pdf

2 See http://www.iraqwarlogs.com/PDF/13/2.pdf

References

Associated Press (AP). (2010). Brits: Investigate Iraq "Abuses". *New York Post*, October 24.
http://www.nypost.com/p/news/international/brits_investigate_iraq_abuses_Ne PNVkZG0iSteHuoG5s8YM

———. (2009). Experts: Torture in War on Terror Eroded Rights. *MSNBC*, February 16.
http://www.msnbc.msn.com/id/29223163/ns/world_news-terrorism/

Benjamin, M. (2007). The CIA's Torture Teachers:Psychologists helped the CIA exploit a secret military program to develop brutal interrogation tactics—likely with the approval of the Bush White House. *Salon*, June 21.
http://www.salon.com/news/feature/2007/06/21/cia_sere/print.html

Bureau of Investigative Journalism (BIJ). (2010a). 76 Cases of Abuse Challenges US report on Iraqi Prisons. *Iraq War Logs*, October 19.
http://www.iraqwarlogs.com/2010/10/19/us-inspectors-clear-iraqi-detention-facilities-but-troops-file-reports-of-horrific-torture/

———. (2010b). Torture Widespread in Iraqi Detention Facilities. *Iraq War Logs*, October 22.
http://www.iraqwarlogs.com/2010/10/22/iraqi-state-torture/

———. (2010c). US Troops Ordered Not to Investigate Iraqi Torture. *Iraq War Logs*, October 22.
http://www.iraqwarlogs.com/2010/10/22/us-troops-ordered-not-to-investigate-iraqi-torture/

———. (2010d). Allegations of Prisoners Abuse by US Troops after Abu Ghraib. *Iraq War Logs*, October 23.
http://www.iraqwarlogs.com/2010/10/23/secret-files-reveal-allegations-of-prisoner-abuse-by-american-troops-after-abu-ghrai/

Bush, G. W. (2004). President's Statement on the U.N. International Day in Support of Victims of Torture. Washington, DC: Office of the Press Secretary, the White House, June 26.
http://georgewbush-whitehouse.archives.gov/news/releases/2004/06/20040626-19.html

CNN. (2009). Top Prosecutor Orders Probe into Interrogations; Obama Shifts Onus. *CNN*, August 24.
http://www.cnn.com/2009/POLITICS/08/24/us.terror.interrogations/index.html

Department of the Army (DoA). (1991). Prisoners of War (Training Circular 27-10-2). Washington, DC: Department of the Army.

http://www.loc.gov/rr/frd/Military_Law/pdf/prisoners-of-war-1991.pdf

Fermino, J. (2010). Wikileak "Treason" Post: Troops and Allies in Peril: Pentagon. *New York Post*, October 23.

http://www.nypost.com/p/news/national/wikileak_treason_post_1puiUzMmR0 ZX6PpAU8RIHK

International Commission of Jurists (ICJ). (2009). Assessing Damage, Urging Action: Report of the Eminent Jurists Panel on Terrorism, Counter-terrorism and Human Rights. Geneva: International Commission of Jurists.

http://ejp.icj.org/IMG/EJP-Report.pdf

Iraq Body Count (IBQ). (2010a). 15,000 Previously Unknown Civilian Deaths Contained in the Iraq War Logs Released by WikiLeaks. *Iraq Body Count*, Press Release 18, October 22.

http://www.iraqbodycount.org/analysis/reference/press-releases/18/

———— . (2010b). Iraq War Logs: What the Numbers Reveal. *Iraq Body Count*, October 23.

http://www.iraqbodycount.org/analysis/numbers/warlogs/

Lowry, R. (2011). Cheney's Victory: Terror War Constants. *New York Post*, January 21.

http://www.nypost.com/p/news/opinion/opedcolumnists/cheney_victory_U5V oa8sc6xEMmef8hI5J2I

Mayer, J. (2008). *The Dark Side: The Inside Story of How the War on Terror Turned into a War on American Ideals*. New York: Doubleday.

Nance, M. (2009). Torture: What did they know, when did they know it? *Foreign Policy*, April 23.

http://experts.foreignpolicy.com/posts/2009/04/23/torture_what_did_they_kn ow_when_did_they_know_it

Newscore. (2010). New Wikileaks Document Suggests U.S Forces Turned a Blind Eye to Iraqi Torture, say media. *New York Post*, October 23.

http://www.nypost.com/p/news/international/wikileaks_may_be_preparing_to_ release_UN9xHfDlNxKFhOvVeT2pVN

The New York Times (NYT). (2010). C.I.A. Interrogations. *The New York Times*, November 9.

http://topics.nytimes.com/top/reference/timestopics/organizations/c/central_in telligence_agency/cia_interrogations/index.html?scp=1&sq=torture%20in%20wa r&st=cse

The New York Times (NYT) (2011). Abu Ghraib [Topic Archive]. *The New York Times*.

http://topics.nytimes.com/topics/news/international/countriesandterritories/ira q/abu_ghraib/index.html

Pither, K. (2008). *Dark Days*. Toronto: Penguin Group Canada.

Record, J. (2003). Bounding the Global War on Terrorism. Carlisle, PA: Strategic Studies Institute, U.S. Army War College.

http://www.globalsecurity.org/military/library/report/2003/record_bounding.p df

United Nations (UN). (1984). Convention against Torture and Other Cruel, Inhuman or Degrading Treatment or Punishment. New York: United Nations

General Assembly.
http://www.hrweb.org/legal/cat.html
WikiLeaks. (2010). War Diary: Iraq War Logs [Database].
http://wikileaks.ch/iraq/diarydig/

CHAPTER FIVE

HUMANIST OR IMPERIALIST? HUMANITARIAN INTERVENTIONISM IN THE POST-COLD WAR ERA

Jessica Cobran

Humanitarianism is a concept bound with controversy, particularly when it takes the form of military intervention and produces casualties. Political leaders in the West, led by the U.S., currently speak as if they feel obliged to engage in the "protection" of civilians, as part of an effort to maintain global "order". Part of the controversy of "humanitarian intervention" stems from the question of when is the right time to intervene (and when is it not). Some will argue that there is never a good time to intervene, while others will argue that it is our duty to protect those who cannot protect themselves. To take the approach of protecting those who cannot protect themselves is dignified in its nature. However, history (and present day events) has shown that humanitarian intervention can also have ulterior motives. What should be kept separate from politics, has become an effective tool to change the politics of any particular "developing" society. Dependent on intervention, access to oils, minerals, and other natural resources along with political access can all be achieved through humanitarian intervention. Humanitarian military intervention has, as I will argue, become a Western way of seeking imperial gain and political advantages at the expense of the so-called developing world. The "new" imperialism has been defined as having several "new features: These are...1) the shift of the main emphasis from rivalry in carving up the world to the struggle against the contraction of the imperialist system; 2) the new role of the U.S. as organizer and leader of the world imperialist system; and 3) the rise of a technology which is international in character" (Magdoff, as quoted in Forte, 2010, p. 5). In this context, there are many ways in which humanitarian interventions are a practice of imperialism.

Humanitarian Intervention vs. Peacekeeping: Defining Concepts

It is imperative to understand and not confuse humanitarian intervention with peacekeeping. Humanitarian intervention involves "the transboundary use of military force and is distinct from crossing borders to provide humanitarian aid, which does not entail military force" (Heinze, 2009, p. 8) and is supposed "to alleviate human suffering" (Heinze, 2009, p. 4). Meanwhile, peacekeeping has to be neutral, impartial, and involves "limited military capability" (Heinze, 2009, pp. 8, 9; Haass, 1999, p. 57). It is virtually impossible to be neutral in those efforts, especially with those that involve fighting among competing "warlords" (Weiss, 2007, p. 77). Although the United Nations (UN) is often associated with peacekeeping missions, UN enforcement operations, on the other hand, are not peacekeeping, but would fall more under "humanitarian intervention" because they are authorized to initiate the use of force to stop "human suffering" (Heinze, 2009, pp. 8, 9). Peace-making is another deceptive term, because it has nothing to do with peace or peacekeeping: generally, "peace-making is used to cover those activities falling between peacekeeping and war-fighting" (Haass, 1999, p. 59).

Some will also define humanitarian intervention by two categories, consensual or imposed. Consensual requires very little firepower, whereas what is being focused on in this essay is *imposed*, i.e. that which requires "heavily armed troops" (Haass, 1999, p. 62) and is the one we have come to know more about in recent years. Humanitarian intervention can involve the deployment of military forces to protect foreign nationals (Finnemore, 2003, p. 53). "'Humanitarian intervention,' the notion that outside parties have the right or even obligation to intervene to help peoples vis-à-vis their own governments or one another," (Haass, 1999, p.12) is something that we have been seeing more of, especially since post-Cold War.

In the post-Cold War era one of the ongoing debates is surrounded by the emphasis of the "obligation to act," (or more recently the Responsibility to Protect [R2P]), including military action (Lepard, 2002, p. 261). Since the end of the Cold War various countries "have intervened militarily to protect citizens" (Finnemore, 2003, p. 52).

Shocked Doctrines

Although the U.S. is not the only Western country to be involved in humanitarian interventions, since President Carter, it has taken on the role based on a narrative where human rights equate to the "soul of American Foreign Policy" (Bricmont, 2006, p. 64). More recently, the Bush Doctrine also included the principle that "the United States must defend liberty and justice because these principles are right and true for all people everywhere" (Coicaud, 2007, p. 141).

Although the definition of humanitarian intervention has been made clear, there are also clear definitions of acceptable grounds for intervention. In the Report of the International Commission on Intervention and State Sovereignty (ICISS) on the Responsibility to Protect, it is made clear in section 4.13 "Responsibility to React," that the intervention must be based on "exceptional circumstances". These "exceptional circumstances must be cases of violence which so genuinely 'shock the conscience of mankind,' [but also] present a clear and present danger to international security, that they require coercive military intervention" (ICISS, 2001, p. 31). Many questions come to mind when hearing this statement. One question is who decides what shocks the conscience of mankind? Is the consciousness of mankind so universal that it is identical across all people and all countries? Is this something that only involves the Western world condemning developing countries, or is it possible that a developing country can intervene in other countries as well? For some homelessness, hunger and a lack of public health care may "shock the conscience of mankind" in the modern day world—if this was to be the case, then we would need to be shocked by the many countries, including Western ones, that have violated basic and fundamental human rights. Also note that section 4.13 says that the humanitarian emergency must also present a clear danger to international security. So what about countries with conflicts that are clearly local? The doctrine also speaks of "*coercive* military intervention". Coercive is an interesting choice of word. Here we can see the distinction between keeping the peace and enforcing one's will on a people. Somalia, Iraq and Afghanistan are just three examples where they have been painted as a danger to international security, giving a reason for why intervention needed to continue or be implemented in those countries. Section 4.13 also describes that the decision to intervene is for extreme cases only, "when all order within a state has broken down or when civil conflict and repression are so violent that civilians are threatened with massacre, genocide or ethnic cleansing on a large scale" (ICISS, 2001, p. 31)—except that such extreme cases have often been either the result of

Western interventions beforehand, or the result of casting certain situations using fanciful interpretations.

There were a few prominent examples of humanitarian interventions that took place in the 1990s.

Somalia

In 1991 Somalia was faced with civil unrest. In 1992 the U.S., under then President George H.W. Bush, led the charge into intervening and providing humanitarian aid. In the earlier stages, the U.S. had military airlifts that were there to secure and provide for a "500-person U.N. force" so that they could deliver aid, such as food (Haass, 1999, p. 43). The mission in Somalia was to have a "short, U.S.-dominated humanitarian phase, which was supposed to create a longer U.N.-dominated political effort" (Haass, 1999, p. 44).

President Bush in his 10 December 1992 address ensured and insisted that the U.S. effort was strictly humanitarian, not strategic (Haass, 1999, p. 44). Although it may have been argued that strides were being made, once the Clinton administration took over in 1993, the involvement in Somalia increased more than was "anticipated" (Haass, 1999, p. 45). Despite the public outcry when several soldiers lost their lives in Somalia, President Clinton sent more troops insisting it had nothing to do with imposing U.S. policy on Somalia, but rather was about "protecting troops and bases, to keep open key roads and ports, to pressure those who would attack the supply routes or U.S. forces, and to provide a context for a Somali political process" (Haass, 1999, p. 46). "A more ambitious set of objectives" was created for Somalia. It became a mission to restore "calm," "disarm warlords," and establishing "credible local police forces in major population centers" (Haass, 1999, p. 45). Somalia all of a sudden became "a threat to international peace and security" (Chesterman, 2002, p. 140). "The new policy [in Somalia] amounted to nothing less than nation-building," (Haass, 1999, 45), a concept that resurfaced in the occupation of Afghanistan and will be discussed further on.

Kosovo

In the late 1990s, NATO made a decision to bomb Yugoslavia when it was accused of trying to ethnically cleanse Albanians living in the region. The example of Kosovo was an "illegal but legitimate" humanitarian intervention (Heinze, 2009, p. 1). Although U.S. military force was against international law, it was said by proponents of the intervention to be "morally justified". One of the reasons why it violated international law was due to the fact that the bombing and intervention in Yugoslavia was not for purposes of NATO's self-defense. In the case of Kosovo, although "some may have had doubts about the legality or wisdom of NATO's action without prior U.N. authorization" numerous Council members still "supported the use of force to protect Kosovo Albanians" (Lepard, 2007, p. 226, Weiss, 2007, p. 79). Perhaps this approval was based on the idea that "human rights are a value far more fundamental than respect for international law" (Bricmont, 2006, p. 61).

Chapter V (The Security Council) of the Charter of the United Nations, Article 24.1 states that "in order to ensure prompt and effective action by the United Nations, its Members confer on the Security Council primary responsibility for the maintenance of international peace and security, and agree that in carrying out its duties under this responsibility the Security Council acts on their behalf" (UN, 1945). Even with this lopsided balance of power in favour of a select few, the U.S., France, and the U.K.—members of the UNSC—violated its provisions by acting without UN approval in Kosovo.

Congo

Lest we also forget about Congo. The atrocities in the Democratic Republic of Congo did not receive nearly as much attention, if at all. In the years falling "between 1998 and 2003 some four million people were killed" without receiving much international attention (Hehir, 2008, p. 113). In the era in which we are living, "there are more resources than ever," but some crises are neglected over others (Weiss, 2007, p. 74). For instance, Yugoslavia received $207 per person in aid, while it was about $8 per person in the Democratic Republic in Congo (Weiss, 2007, p 74). Mahmood Mamdani asks the question, "could the reason be that in the case of Congo, Hema and Lendu militias—many of them no more than child soldiers—were trained by America's allies in the region, Rwanda and Uganda?" (2009, p. 63).

What is the Problem with Humanitarian Intervention?

As citizens of so-called "first world countries" we are led to believe that we have a special "Responsibility to Protect". Tony Blair, former British Prime Minister, used arguments similar to those that would later be codified in the R2P doctrine to explain the invasion of Iraq: "we surely have a responsibility to act when a nation's people are subjected to a regime such as Saddam's" (Weiss, 2007, p. 124). It is hard to sit back and watch human suffering and the submission of basic human rights through our various media outlets. As citizens of a privileged country some believe that it is our right—no, our *duty*—to protect those who are oppressed (elsewhere). The ideology of humanitarianism is a beautiful concept. So what is the problem with it; why even debate it? The issue is that humanitarian and military interventionism goes beyond just protecting people.

It has been argued that "no public policy issue is more controversial than the use of military force. As U.S. experience in Somalia and Bosnia in the early 1990s showed, it matters not whether we choose to intervene or stay aloof; the debate can be equally heated" (Haass, 1999, p. 1). But what if we do choose to intervene? What are the consequences at hand?

One of those consequences can be increased bloodshed. The question that we really have to ask ourselves is does humanitarian intervention do more harm than good? Gandhi once was quoted as saying that, "I assert in all humility, but with all the strength at my command, that liberty and democracy become unholy when their hands are dyed with innocent blood" (Chesterman, 2002, p. 88), hence even *if* the intention was good, the end doesn't always justify the means. This sentiment was again repeated by Former Malaysian Prime Minister Mohamed Mahathir in his belief that "the fanaticism of champions of human rights have led to more people being deprived of their rights, and many of their lives, than the number saved [...] Just as many wrong things are done in the name of democracy and human rights" (Bricmont, 2006, pps. 82, 83). This proved to be the case in Somalia, because during the humanitarian intervention eventually what ensued was an increase in violence, both the lost lives of U.S. troops and of course many civilian deaths (Haass, 1999, p. 45).

Power: In the Hands of Few and Far Between

In the centuries of formal colonialism, the main powers were Britain, Spain and France among others. They jousted for positioning by grabbing up as much territory and resources as possible. In the contemporary period, we have witnessed some change in terms of which is the leading power. Despite "economic downturns" the U.S. is still the leading superpower, at least in political and military terms. Some have argued that this amount of power translates to "the United States [being able to] define the terms of the debates and hold the keys—even for issues about which it cares marginally" (Coicaud, 2007, p. 134). During Bill Clinton's presidency, "French Foreign minister Hubert Vedrine coined this as '*hyperpuissance*' (hyperpower) to characterize the inescapable reality of American political, economic, and military dominance of the world" (Heinze, 2009, p. 127). Although critics may say that "many people outside the United States simply do not trust America to use its enormous power wisely or well" (Heinze, 2009, p. 127), but there are other countries such as France and the United Kingdom (both are also part of NATO) who play major roles in humanitarian interventions as well.

In the issue of Rwanda, the world took too long to respond, as we are often told. Many years later reports were published laying blame on the UN and its security council for failing the Rwandan people during the 1994 genocide (Lepard, 2002, p. 261). Again, the questions arise, what makes the UN and U.S. jump into certain conflicts and not others? Some countries seem to require careful deliberation, while in other cases there is a rush to intervene. With the increased awareness on international terrorism, this has also slightly changed the outlook of interventions. Although there were massive doubts about the possibility of Weapons of Massive Destruction (WMD), some allies of America joined in the fight against Iraq, even if reluctantly. In fact, even though no WMDs were found in Iraq, America is still there on the basis of "stabilizing Iraq" and "constructing democracy" (Bricmont, 2006, p. 64).

Examples of this so-called "stabilization" can also be found in Somalia in 1991, Haiti in 1994, and Bosnia in 1995 (Heinze, 2009, p. 9). The 1990s have become known for an era of humanitarian intervention (Haass, 1999, p. 20), which continues into the present.

The 1990s

The large number of civil wars in the 1990s "challenged" traditional humanitarianism, which involved "consent, impartiality, and neutrality" (Weiss, 2007, pps. 73, 142). Many challenges ensued in the 1990s even though the "post-Cold War world promised a kinder, gentler 1990s" (Weiss, 2007, p. 142) and "by the onset of the 1990s, it looked as though a world was emerging in which democratic and market-oriented governments would dominate, in which age-old conflicts were being solved, and in which the United Nations was finally beginning to resemble the institution desired by its founders" (Haass, 1999, p. 2). However, this was not the case. The 1990s saw numerous civil conflicts that came to the forefront and pushed the limits of humanitarian interventions, especially on the part of the U.S. It was an era that "led to an entirely new global security environment, marked by a focus on internal rather than inter-state wars" (UN, 2011, ¶ 1). More and more questions began to arise of when and when not to intervene. For instance, without the permission of the UN, the U.S. stepped into the conflicts in Kosovo, taking a more unilateral approach, yet the U.S. sat by as a massacre of people, genocide, ensued in Rwanda. Some have argued that intervention is really a matter of strategy more than it is about helping those in need, yet it seems like a *Disney Damsel-in-Distress Syndrome* because it's just a bunch of caring people going over to (some) desperate "third-world" countries to help rescue them. Just as in Disney films there is a victim who needs rescuing by a brave, sweet prince; the West is often portrayed as the big, strong brother who can go into other countries to protect the world from the threat to democracy, freedom, and human rights.

Humanitarianism: A Means to an End

In this game of humanitarian intervention, alliances, strategy and interests play a pivotal role in interventions, or lack thereof. For example, in 1999, the East Timorese people voted for independence from Indonesia. In retaliation violence ensued against the East Timorese by the Indonesian army, on a scale that was genocidal. The UN Security Council failed to respond without the consent of the Indonesian government (Lepard, 2002, p. 24). Criticism later concluded that "it would, therefore, seem logical to conclude that the unwillingness to intervene without Indonesia's consent derived from a desire to avoid a total breakdown in relations with a strategic ally" (Hehir, 2008, p. 79). When other goals or interests are at stake, there is a reluctance to commit to "humanitarian action" (Finnemore, 2003, p. 65).

With the West's ties to countries like Indonesia, "it is additionally clear that Western states have no compunction about cultivating alliances with states with poor human rights records" (Hehir, 2008, p. 80). For instance, the U.S. has close relationships with Saudi Arabia and others who are "noted as systematic violators of human rights" which "implies that key strategic considerations are given preference over humanitarian concerns" (Hehir, 2008, p. 80). However, America is not the only country that behaves in this manner. The French were lenient towards Saddam Hussein and Iraq, with "long-standing support of his secular regime" and the French also had economic interests in Iraq, predominantly their obvious dependence on Iraq's oil (Coicaud, 2007, p. 154).

There is an ally-enemy divide—if you are not with the U.S., you are against it. The U.S. plays a leading role in determining which cases deserve "humanitarian" intervention (Coicaud, 2007, p. 141). The U.S. has also played a role in financing foreign leaders from 1948 to 2005 in over 34 countries (Bricmont, 2006, p. 85). The way in which U.S. national security interests have been defined leads to neutralizing opponents to allegedly ensure its own security (Coicaud, 2007, p. 141). Bricmont writes that "foreign aid budgets amount to a tiny fraction of GNP. And even less if we subtract the share that goes to military cooperation or promotion of our own business interests" (2006, p. 80).

In the case of Darfur, the West's initial refusal to intervene could also suggest that deciding when to intervene remains influenced by strategic interests more than by the "scale of [any] suffering taking place" (Hehir 2008, p. 79). During Darfur and the unfolding of the Iraq war, "U.S. intentions became increasingly suspect [...and prompted] further suspicion that the United States was essentially after Iraq's oil and waging an imperialistic war against Arabs and Muslims" (Heinze, 2009, p. 127). During the most recent war with Iraq "neo-conservative policymakers" with the "Project for a New American Century" stated that the Iraqi invasion was based more on protecting "vital interests" in the Gulf, than it was for human rights (Bricmont, 2006, p. 63).

The declaration of genocide in Darfur (by U.S. President George W. Bush) also prompted "accusations that the United States was essentially hyping the charge of genocide as a smokescreen behind which it could invade Sudan for other reasons, such as access to the vast oil reserves quite obviously coveted by U.S. oil companies" (Heinze, 2009, p. 128). These suspicions also grew after the fact that according to Jan Pronk, a U.N. Special Representative in Sudan, "the mortality and malnutrition rates had decreased dramatically in 2005" (Mamdani, 2009, p. 49). In 2007, Darfur and Chad also became listed on the U.S. Pan Sahel Initiative on counterterrorism; coincidentally both places in Sudan are rich in oil and possibly minerals (such as uranium) (Mamdani, 2009, p. 55).

Condoleeza Rice, who was the National Security Adviser during George W. Bush's first presidential term, said in an issue of *Foreign Affairs* that "Humanitarian problems are rarely only humanitarian problems; the taking of life or withholding of food is almost always a political act" (Coicaud, 2007, p. 147). Thus even Bush's senior foreign policy advisor candidly placed humanitarianism on a political plane, in spite of the rhetoric that tried to enforce a distinction between humanitarianism and politics.

Although perhaps humanitarian intervention should be separate from politics, it evidently is not (Weiss, 2007, p. 77). In the 1990s "Security Council authorization became politicized and was not based on an adherence to legal provisions contained in [Chapter VII of the UN] Charter" (Hehir, 2008, p. 18). Humanitarian intervention is beyond terminating an emergency and saving lives in the present, it involves changing the "structural conditions" of the society (Weiss, 2007, p. 82), showing the political advantages that can come with humanitarian interventions.

Unilateralism

"While playing lip service to the United Nations and multilateralism, the Bush administration" showed through their actions that they were not going to be constrained by them (Coicaud, 2007, p. 148). Iraq was a great demonstration of this refusal to be bounded by the UN and multilateralism (Coicaud, 2007, p.150). Multilateralism has usually been used in a way to "legitimize action by signaling broad support for the actor's goals" (Finnemore, 2003, p. 82). However, there have been several cases in both the past and the present where the U.S. (and NATO) went along in a more unilateral way, without the consent of the UN. Clearly the issue of when to act is not dependent upon multilateral support, although it is actively sought when it seems attainable, and is then used to justify intervention. In other cases, it is simply dismissed.

Nation-Building

We often see the promotion of democracy and human rights (Weiss, 2007, p. 82) and nation-building. Nation-building is "an extremely intrusive form of intervention, one that seeks to bring about political leadership" (Haass, 1999, p. 61) and create institutions and a society which are different from those that currently exist in the particular country. For the U.S., nation-building is about creating a democracy (ostensibly) with "free-market practices" (Haass, 1999, p. 61). However, this type of system is not suited for everyone. For example, introducing Cuba to such "democracy" may run the risk of a capitalist transformation of the economy with IMF requirements, resulting in the abolition of free health care for the multitude (Bricmont, 2006, p. 85). Although some may argue that democratizing places such as Cuba would "free the people," they may also lose their access to other rights that citizens in the Western world do not have.

No Such Thing as Impartiality

The UN Security Council has exemplified double standards (Weiss, 2007, p. 123) and so has the U.S and NATO when it comes to interventions. To say that humanitarian interventions are completely impartial would be having blinders on, because "all interventions are prompted by a mixture of motivations in some way" (Finnemore, 2003, p. 56).

Article 55 (Chapter IX: International Economic and Social Co-Operation) of the Charter of the UN states:

> "With a view to the creation of conditions of stability and well-being which are necessary for peaceful and friendly relations among nations based on respect for the principle of equal rights and self-determination of peoples, the United Nations shall promote: a) higher standards of living, full employment, and conditions of economic and social progress and development; b) solutions of international economic, social, health, and related problems; and international cultural and educational cooperation; and c) universal respect for, and observance of, human rights and fundamental freedoms for all without distinction as to race, sex, language, or religion". (UN, 1945)

It would seem that many countries are in violation of this, but which ones are subject to intervention for violating these rights is not nearly as inclusive a list. In the end, waging a war to fight for peace and human rights is an oxymoron and "nothing in U.S. policy indicates the slightest sincere concern for human rights and democracy" (Bricmont, 2006, p. 67).

Appealing to Human Sensibilities through Mainstream Media

In 1992, Nancy Kassebaum and Paul Simon wrote for *The New York Times* an article entitled, "Save Somalia from Itself". The opening paragraph reads:

> "Since civil war broke out in Somalia on Nov. 17, as many as 20,000 people, mainly civilians, have been killed or wounded. Many more Somalis are starving to death because the fighting is preventing relief agencies from distributing emergency food supplies in the East African country. Most of the thousands of injured Somalis cannot make it to the few rudimentary hospitals remaining in Mogadishu, the capital. The war continues to escalate, yet the world largely ignores it". (Kassebaum & Simon, 1992, ¶ 1)

This is merely one example of how the media are used to convey cues to us about what our appropriate emotional responses should be to other nations' problems.

Intervening in other countries involves appealing to the public's emotions in order to receive approval. We can see many examples of this (9/11, the War on Terrorism, Save Darfur etc.), and the media is the perfect tool to exploit these emotions. Non-Governmental Organizations (or NGOs) understand that identifying with the victim and creating empathy for the victim is vital to whatever the cause may be (Finnemore, 2003, pps. 157, 158). The largest NGOs have grown and so have the "size of the NGO 'humanity footprint' [showing] a 91 percent growth from 1997 to 2005" (Weiss, 2007, p. 148).

In order for the emotional responses to successfully arise, there has to be "manipulation through exposure" (Finnemore, 2003, p. 157). The media and NGOs have worked well together to create "familiarity where little existed" before the media exposure began (Finnemore, 2003, p. 158). How does this relate to humanitarian interventions? Quite simply, "by manipulating empathy, agents can change perceptions about what kind of situation exists and whether it requires military force" (Finnemore, 2003, p. 158).

The reasons behind invading Iraq changed several times, until it became a humanitarian issue. Suddenly the "U.S. administration emphasized the humanitarian argument for the invasion, essentially arguing that the war was justified because it removed a tyrant and was bringing freedom and democracy to Iraqis" (Heinze, 2009, p. 127). This argument had received a lot of skepticism by scholars of humanitarian intervention, raising fears "about how the humanitarian argument could be used as a pretext to mask the exercise of hegemonic power" (Heinze, 2009, p. viii). Yet despite the skepticism, people followed along and watched a war ensue that still continues on to this very day. Whatever the U.S. Administration said, the media or so-called "free press" followed suit as it went from trying to find WMD to trying to create a free and democratic society for the oppressed people of Iraq under the Saddam regime.

Although the media in the U.S. is called the free press, there is little that is "free" about it. When it comes to foreign policy, "the 'free press' is remarkably uniform," but by being called the free press, it "makes it a more efficient propaganda tool" (Bricmont, 2006, p. 69). This is because when the media is directly censored by the state, there is an automatic distrust of the source, whereas calling it a "free press" creates the impression that it is an unbiased and reliable resource (Bricmont, 2006, p. 69).

Conclusions

In a utopian world, humanity would reign. It would be a place where people love one another; equity and equality would be spread across the world. There would be no such thing as genocide, famine and malnutrition. However, this is not the type of world we live in. We live in a global society where the gap between the rich and the poor is widening in most places. This makes it hard for anyone to have the moral justifications to play the impartial judge and jury.

Although today's version of humanitarianism involves a multilateral approach, to create more legitimacy, the U.S. and NATO on several occasions have intervened without seeking the permission of the UN. The U.S., leading the charge in politics and power across the world is able to make these types of violations against international law. They can also influence others to follow them in their decisions to intervene (or not to intervene).

In the 1990s, there were many instances where the U.S. had the option to intervene or stand by and watch idly. Kosovo, Haiti, Somalia, Congo, and Rwanda are just a few of the places where the U.S. had some type influence on whether or not intervention would ensue. Darfur, Iraq, and Afghanistan having been placed at the forefront in the new millennium, are places that U.S. has had motivations for intervention that went beyond any purported humanitarianism. The decision to intervene was based on strategic motivations.

Although it may seem as though humanitarian intentions are pure, when the U.S. (and the West) or even UN Security Council step in to intervene, one should never forget that there is always something to gain by them when intervening. Even the fact that they choose not to intervene in countries such as Saudi Arabia, exemplifies the fact that it is beneficial for countries who violate human rights, to be an ally to a political powerhouse, such as the U.S. Key resources such as oil and an opportunity to exert political influence have been connected to the U.S. and NATO interventions.

The NGOs and the media also play a powerful role in the agendas of the West. The "free press" is an effective tool used to promote the agenda of the West. It is an effective way to implement the views that the government wants its people to see—even if the press tries to portray the fact that they are unbiased and separate from the government.

As stated from the onset, humanitarian intervention is a very controversial topic. No matter what side you fall on, it is important to stay well-informed. Yet, the twists and turns of the portrayals of interventions are too blatant to ignore. From intervening to find weapons that can disrupt international security, to all of sudden becoming a mission to save the people who are being oppressed, there is one common dominator—an opportunity to imperialize.

References

Bricmont, J. (2006) *Humanitarian Imperialism: Using Human Rights to Sell War*. New York: Monthly Review Press.

Chesterman, S. (2002). *Just War or Just Peace?* New York: Oxford University Press.

Coicaud, J. (2007). *Beyond the National Interest: The Future of UN Peacekeeping and Multilateralism in an Era of U.S. Primacy*. Washington: United States Institute of Peace.

Finnemore, M. (2003). *The Purpose of Intervention*. New York: Cornell University Press.

Forte, M.C. (2010). Introduction: The "New" Imperialism of Militarization, Humanitarianism, and Occupation. In M.C. Forte (Ed.), *The New Imperialism, Vol. 1: Militarism, Humanism, and Occupation* (pp. 1-29). Montreal, QC: Alert Press.

Haass, R.N. (1999). *Intervention: The Use of American Military Force in the Post-Cold War World (Rev. Ed.)*. Washington: Brookings Institution Press.

Hehir, A. (2008). *Humanitarian Intervention after Kosovo: Iraq, Darfur and the Record of Global Civil Society*. London: Palgrave MacMillan.

Heinze, E.A. (2009). *Waging Humanitarian War*. Albany: State University of New York Press.

International Commission on International and State Sovereignty (ICISS) (2001). *The Responsibility to Protect*. Ottawa: International Development Research Centre.

Kassebaum, N., & Simon, P. (1992). Save Somalia From Itself. *The New York Times*, January 2. http://www.nytimes.com/1992/01/02/opinion/save-somalia-from-itself.html

Lepard, B.D. (2002). *Rethinking Humanitarian Intervention*. University Park, PA: The Pennsylvania State University Press.

Mamdani, M. (2009). *Saviors and Survivors: Darfur, Politics, and the War on Terror*. New York: Pantheon Books.

Weiss, T.G. (2007). *Humanitarian Intervention*. Cambridge: Polity Press.

United Nations (UN). (1945). *Charter of the United Nations*. New York: General Assembly of the United Nations. http://www.un.org/en/documents/charter/

———. (2011). Terrorism. http://www.un.org/en/globalissues/terrorism/

CHAPTER SIX

TRANSPARENCY SHIFT:
AN OVERVIEW OF THE REALITY OF WIKILEAKS

MacLean Hawley

In 2010, WikiLeaks became prominent in the general public's eye when it released the "Collateral Murder" video and subsequently came fully into view with the stunning leak of a quarter of a million U.S. diplomatic cables and other war records. Whether "pro" or "anti" such disclosure, people around the world were concerned with how WikiLeaks would affect the way the U.S. functioned internationally. During this time, I thought that arguments "for" or "against" WikiLeaks were irrelevant, since it was debatable WikiLeaks would have any impact at all, let alone global change for the positive or negative. Now, roughly six months later, evidence that WikiLeaks is changing, and maybe even damaging, the American Empire (and changing the world itself) is accumulating. This evidence is proof that a small group of dedicated people can change the course of history. We now have answers to the question above and to many others, pervasive since the creation of WikiLeaks. I find that these questions usually took the following forms: Has WikiLeaks really been "opening governments" as their slogan states? Or are writers such as Jaron Lanier (2010) correct when they say that WikiLeaks has accomplished the antithesis of their avowed goal, working instead to close governments? To put it bluntly, has WikiLeaks in its five short years, actually managed to change anything at all?

In order to answer questions like these, we first must understand what the organization actually claims to be doing. The WikiLeaks website calls itself a "secure and anonymous way for sources to leak information," working to fulfill its goal of bringing "important news and information to the public" (WikiLeaks, 2011, ¶ 1). The organization states that publishing this information "improves transparency, and this transparency creates a better society for all people" (WikiLeaks, 2011, ¶ 7). Though a seemingly grandiose statement, this makes sense when scrutinized. A more transparent government will in theory better inform the public. A more informed public has more power and control. As John F. Kennedy said, "The ignorance of one voter in a democracy impairs the security of all" (Kennedy, 1963, ¶ 16).

Figure 6.1

"We Open Governments". This is one of a number of popular posters that came out in late 2010, produced by anonymous WikiLeaks fans, and distributed in the public domain. This poster evokes "Top Secret America" with the Pentagon building opened to leaks and thus public inspection of its interior.

But this is an awful lot for one website to promise. Can an organization just shy of its fifth year create any change, not to mention promising change that will affect the whole scheme of global politics, and bring power to the people? If any change is made, is it necessarily a good thing?

These are tough questions, which first require a more in-depth understanding of WikiLeaks. Throughout this organization's history there have been some changes that have incurred criticisms from skeptics, and even caused loyal followers to sour.

Change within the Beast

"Any change, even a change for the better, is always accompanied by drawbacks and discomforts".–Arnold Bennett (Quoted in Peter, 2010, p. 115)

Although WikiLeaks' cofounder Julian Assange may be hesitant to admit it, there has been a substantial amount of change during WikiLeaks' short history. Today, the way WikiLeaks functions resembles a news website more than anything else. All of the documents donated to them are summarized and analyzed by their employees, with no input from the general public. This is a far cry from what the creators originally had in mind. When WikiLeaks was founded in October 2006, the official documents that were leaked to them were posted with editable analyses and summaries (the assumption being that visitors would summarize and analyze them in a manner similar to how visitors to Wikipedia edit the material they find there). Unfortunately, when this failed to happen, WikiLeaks took more control over the material they posted, and retracted user-editing functions. Julian Assange blames this change on the public:

"Surely if you give all of those people working on all the junk on Wikipedia...surely those people will step forward given fresh source material and do something. No, it's all bullshit. All bullshit. In fact people write about things in general, that aren't part of their career, because they want to display their values to their peers. Actually, they don't give a fuck about the material". (Berkeley Graduate School of Journalism, 2010, 36:34)

Some might argue that WikiLeaks never was, and never claimed to be an actual Wiki. The one thing it had in common with something like Wikipedia was that it was created with the same shareware as other Wikis. Interestingly, WikiLeaks never publically acknowledged that they were using this piece of shareware, which in turn violates its terms of use, according to one critic (Lamo, 2010, ¶ 5). The reasons for Assange's contradictory denials of change are unclear, but what is clear is that many are highly critical of the website's shift away from what the "wiki" in its name implies. Even Adrian Lamo, a hacker and once a self-declared supporter of WikiLeaks, claims that the website should not have "wiki" in its name at all. Not only does he believe that the name is misleading, but he also speculates that the only reason why the "wiki" was included was to siphon some success off the popular website Wikipedia (Lamo, 2010, ¶ 11).

Curiously enough, Julian Assange has completely denied any change whatsoever in WikiLeaks. After being asked why the shift from user-edited to staff-edited summaries and analyses occurred, Assange responded, "No, there hasn't been a change, whatsoever" (Berkeley Graduate School of Journalism, 2010, 36:13) and, "That's part of the rightwing reality distortion field" (Berkeley Graduate School of Journalism, 2010, 35:55). Even though he made these bold claims, Assange later goes on to contradict himself by blaming the change on the laziness of WikiLeaks visitors.

Even if WikiLeaks is growing further removed from what the "wiki" in its name implies, its overall goals have remained largely the same. Ultimately this change has been to make these goals more accessible, even if it means they meander a bit from their original form.

Figure 6.2

Julian Assange, WikiLeaks founder. (Photograph by Carmen Valino, made available under a Creative Commons License.)

Changes in the Media

"Technology will make it increasingly difficult for the state to control the information its people receive....The Goliath of totalitarianism will be brought down by the David of the microchip".—Ronald Reagan (Reuters, 1989, ¶ 7)

Augmented Mediascapes

WikiLeaks is in no way the first organization to publish leaks. From Daniel Ellsberg's leaking of Pentagon documents to *The New York Times* in 1971, to Jeffrey Wigland creating the so-called "tobacco-gate" by leaking information about the dangers of tobacco to CBS in 1996, it is clear that news organizations are both savvy and avid users of information leaks. What is different with WikiLeaks, however, is the new sort of "middleman" role that WikiLeaks plays. Not only does WikiLeaks provide a safe method for leakers, it is also the only "free" news organization out there (Shirky, 2011, ¶ 10).

Pre-WikiLeaks, all major news networks were tied to one country or another. Even multinational news corporations like Al-Jazeera or BBC had ties to their respective home countries, which made publishing their respective governments' "sensitive documents" a bit tricky. If publishing national secrets was technically legal in a host country, news organizations still are not likely to publish them since the organizations themselves are often tied closely to government aid (Shirky, 2011, ¶ 10). WikiLeaks is different; it has no country. It is the first truly international, truly Internet-centric news organization. This lack of national allegiance allows WikiLeaks to publish material that would otherwise be non-publishable. Its transnational presence also allows it to take advantage of multiple nations' security laws by cleverly routing the flow of all their information, which further adds to its impenetrability. As put by Internet specialist Clay Shirky (2011, ¶ 9), "WikiLeaks is headquartered on the web; there is no one set of national laws that can be brought to bear on it, nor is there any one national regime that can shut it down".

Although WikiLeaks was the first to start the truly nation-less trend, other news providers have since jumped on board. Al-Jazeera and *The Guardian* used similar transnational information routing techniques to ensure security when they released the Palestine Papers. WikiLeaks has also inspired "copycat" leak websites. OpenLeaks was one of the first of these, designed by Daniel Domscheit-Berg as a "WikiLeaks 2.0," but presumably in competition against WikiLeaks. Other sites, such as QuebecLeaks and BalkanLeaks, have been created to deal with local issues. Certainly, without WikiLeaks to inspire them, news corporations would not have sought to enter the field of transnational leaks, and communities would not have had an example to base their local leak sites upon. While it is possible that these changes would have eventually occurred without WikiLeaks to lead the way, it seems more than probable that WikiLeaks has been the catalyst. Shirky believes that WikiLeaks was solely responsible for this shift, and stated that WikiLeaks "isn't just a new entrant in the existing media landscape. Its arrival creates a new landscape" (Shirky, 2011, ¶ 3)

Showing the Blood and Rust

One side effect of WikiLeaks' new media landscape is the perspective of war offered to the American public, a perspective that had been obscured since the end of the Cold War. In modern times, American opinion has turned war into something completely different from what it was, and from what it, in reality, still is. Historian and writer Andrew Bacevich describes this idea of high-tech/modern warfare as such:

> "by the turn of the twenty-first century, a new image of war had emerged...war was becoming surgical, frictionless, postmodern, even abstract or virtual. It was 'coercive diplomacy'—the object of the exercise no longer to kill but persuade. By the end of the twentieth century Michael Ignatieff of Harvard University concluded, war had become 'a spectacle'. It had transformed into some kind of spectator sport". (Bacevich, 2005, p. 20)

This new portrait of modern war as frictionless, and in a sense clean, is in stark contrast to war's previous identity. Bacevich, eloquently again, describes public opinion prior to the end of the Cold War: "the modern battlefield was a slaughterhouse, and modern war an orgy of destruction that devoured guilty and innocent alike" (2005, p. 20). One of the main reasons for this transition was the military's adoption of technology, which led the public to believe that war would also adopt all the clean precision of the new technology. Facilitating this is the growing number of senior officers "waging bureaucratic warfare, [and] manipulating the media" (Bacevich, 2005, p. 30). Bacevich provides a useful summary, "by the dawn of the twenty-first century the reigning postulates of technology-as-panacea had knocked away much of the accumulated blood-rust sullying war's reputation" (Bacevich, 2005, p. 22). This sanitization of war's image took place over about 40 years, but WikiLeaks in its five short years has managed to remind the public that war is just as bloody as it ever has been.

This brings us to one of WikiLeaks' most famous leaks, 2010's Collateral Murder (WikiLeaks, 2010). This leak comprised video footage from a U.S. Apache helicopter showing Marines killing Iraqi civilians, and specifically Reuters journalists and children. It was particularly disturbing since at points throughout the video the Marines can be heard laughing as they shoot from their helicopter. This video became viral on the Internet, with over 10 million views on YouTube alone. Close on the heels of this leak, WikiLeaks released The Afghan War Diaries, which was a cache of over 90,000 U.S. intelligence reports on the war in Afghanistan. This leak was widely received by the public, and provided "a devastating portrait of the failing war in Afghanistan" (Davis & Leigh, 2010, ¶ 1).

The wide public impact of this video and report caused an almost unprecedented amount of discussion on the subject of America's two wars, discussion that led to an increase in battlefield media coverage (which was almost absent pre-2010). By revealing the unglamorous nature of war, WikiLeaks helped nudge the U.S. public approval of war downwards. In 2009, 51% of the U.S. public approved of the war in Afghanistan, but in 2010, after The Afghan Diaries and Collateral Murder, the public sharply condemned the war, with only 14% approving (Elections Meter, 2011). Although WikiLeaks may not have been the only factor to influence the public's opinion of the wars, these popular leaks have certainly helped. Without these leaks, it is not only arguable that the U.S. population would still have a more positive opinion of these wars, but also of war in general. Indeed, without WikiLeaks the U.S. may well have used the revolutions in the Middle East as an opportunity to more forcefully and directly expand their empire.

Transparent Haven

WikiLeaks may be changing the image of war in America, but there is one country that has been actively shaping itself in WikiLeaks' image: Iceland. In December of 2009, WikiLeaks came before Icelandic Member of Parliament Birgitta Jónsdóttir with the precursor to the Icelandic Modern Media Initiative (IMMI). This law was designed to make Iceland into a free press haven, explained by Jónsdóttir as the "reverse ideology of tax haven, where they pick good legislation from around the world to create secrecy, we want to pick the best legislation to create transparency" (Berkeley Graduate School of Journalism, 2010, 3:20). After fine-tuning by Jónsdóttir, the IMMI was introduced in February of

2010, and unanimously adopted on June 16th of the same year. Not only did passing this law create new legal standings protecting freedom of information, but it also required 13 existing laws to be edited to further support this freedom. Although the law has passed, it will take more then a year for all of the other laws to be edited, and IMMI to be fully put into play.

In this case, WikiLeaks was in the right place at the right time (and with the right people). One of the reasons the law passed with so few restraints was the 2008 financial meltdown in Iceland, a by-product of poor legislation. This led to Iceland's realization that legislative overhaul was necessary, which facilitated the passage of these new laws (Berkeley Graduate School of Journalism, 2010, 5:19). Jónsdóttir was one of many Icelanders to witness this, but she was the only one who was both a WikiLeaks volunteer and a Member of Parliament. In this case, everything was set up for a reform to happen in Iceland, and WikiLeaks took advantage of a good opportunity.

Thanks to WikiLeaks, Iceland is the first, and only, country to have modern legislation regarding the press and the Internet. The amount of change that WikiLeaks brought about in Iceland is surely immense, but what is yet to be seen is how Iceland will influence WikiLeaks. It seems probable that the "Transparency Haven" would be the perfect home for WikiLeaks, and Julian Assange himself has said that the creation of IMMI is "likely to encourage the international press and Internet start-ups to locate their services here" (Vallance, 2010, ¶ 8). Though it is already being used in a transnational sense (information routing), it is very possible that WikiLeaks might see benefit in becoming a national organization.

Change in Africa

> "The number one benefit of information technology is that it empowers people to do what they want to do" (Ballmer, 2005, ¶ 2).

Corruption and Kenya

In Iceland, WikiLeaks helped a government in its decision to become more transparent. However, most of the time governments have no say in whether or not WikiLeaks will affect them, or how. The first major change that directly occurred from a WikiLeaks leak happened in 2007, in Kenya. In order to fully understand what WikiLeaks accomplished, a brief back-story of Kenyan presidential elections is necessary.

In 2002, after Daniel Arap Moi stepped down from his 24-year rule as president, Kenya's fledgling democracy had one its few successful presidential elections, and elected Mwai Kibaki. During the time of the election, Kibaki and a group of supporters commissioned the Kroll Report, which documented that the former president was guilty of secretly siphoning off over $2 billion from the Kenyan government. Kibaki used this report as political leverage within the government, but after the election he chose to keep it secret.

In 2007, during Kenya's next national election, the report became, as Julian Assange put it, a "dead albatross around Kibaki's neck" (Anderson, 2010, 3:30). This was because in an attempt to better the campaign for his second term, Kibaki promised Moi protection and safety from his illegal actions when he was president in exchange for money, and votes from the Kenyan tribes that were still loyal to Moi. Since Kibaki's government had been keeping this report secret, the Kenyan public was unaware of the full extent of their former president's crimes.

Three days after Kibaki teamed up with Moi, WikiLeaks leaked the Kroll report. According to a Kenyan intelligence report, the association of Moi's corruption with Kibaki shifted his votes in the general election by 10%. This shift was more than enough to give his opponent, Raila Odinga, the majority, but paradoxically, Kibaki still won the 2007 election. It seems pretty clear that election was illegitimate, and one Western ambassador claimed, "This was rigged" (Gettleman, 2007, ¶ 18).

In releasing this leak, WikiLeaks better informed the public, which empowered them to make very different decisions than they would have without this information. Kibaki's own corruption was also demonstrated when he rigged the 2007 elections to compensate for the public's shift. Although Kibaki is still in power, WikiLeaks has filled its grand promises of empowering people by informing them, even though the people did not quite get what they wanted.

Tunisia and Assange

With revolutions occurring all across the Arab world, the public and the media (White, 2011) often point to WikiLeaks in their search for a cause. This is most visible in the revolution that occurred in Tunisia.

After a little more then a month of intense protests, Tunisian president Zine El Abidine Ben Ali fled Tunisia on 11 January 2011. While the end result of these protests is still not clear, the removal of a dictator seems like a step in the right direction. With skyrocketing food prices, a lack of freedom of speech, and poor living conditions, there is no question why these protests occurred. What is debatable, however, is what actually sparked them. Julian Assange seems to thinks that it was WikiLeaks, and its release of diplomatic cables showing the corruption of the Ben Ali regime. In late January of 2011 when Assange was discussing the power that WikiLeaks has, he said that there are documents that are "important to get out to the public in a responsible manner that have the potential for great change—for example, this recent revolution in Tunisia" (Kroft, 2011, 22:14, 7:15). Assange often makes grandiose claims like this, but there is, no matter how stretched, undoubtedly there is some truth in them. However we must remember that among Tunisians, the spark that started the protests is generally considered to be the suicide of Mohammed Bouazizi, who lit himself on fire in order to draw the public's attention to government corruption.

Although WikiLeaks was not the "spark" that some claimed it to be, there is evidence that suggests WikiLeaks did play a role in "fueling" the protests. Not only did WikiLeaks' leaked cables solidify popular beliefs of extreme government corruption, but according to Assange they also showed that if it came down to a fight between the military on the one hand, and President Zine al-Abidine Ben Ali's political regime on the other hand, the U.S. would probably support the military (BBC, 2011). This could have bolstered public participation in the protests, and swayed the opinions of those who still believed Ben Ali or feared that by joining the protests they would be opposed by the U.S. Although WikiLeaks did not start the riots, and this is not the fabled "WikiLeaks revolution," the organization is due credit for its support to the protests. WikiLeaks acted like a catalyst, and fanned the flame of change. Once again, supplying the public with information to facilitate its own decision-making is WikiLeaks' goal, and here it was achieved.

The Egyptian Uprising

Even if the role WikiLeaks played in the Tunisian revolution was not as grand as some might have thought, the effects of this revolution go beyond just one country. It is popularly believed that the revolution in Tunisia inspired the people of Egypt to do the same. Parallels do exist between the two revolutions. Corrupt government, food prices, and joblessness are just a few of the shared issues. The revolution in Egypt began a month after the revolution in Tunisia, and according to Julian Assange, "there's no doubt that Tunisia was THE example for Egypt...and all the protests that have happened there" (Davis, 2011, 45:00) which by his logic would make WikiLeaks the start of the whole trend. Saying that it was "THE example", although it could be technically true, seems to give too much credit to the wrong party. The Egyptian revolution was a long time in coming, even longer than the Tunisian revolution. President Ben Ali ruled Tunisia for 23 years, whereas President Mubarak ruled Egypt for almost 30. Tension against

Mubarak was so high before the protests that Egyptian citizens would commonly curse his name in public, which was for many years a punishable offence.

Again, WikiLeaks claims are somewhat exaggerated, but not false. Just as WikiLeaks "fanned the fire" in Tunisia by releasing cables that incriminated the government, it attempted to do the same in Egypt. After three days of protests, WikiLeaks released cables that showed further evidence of Egypt's police brutality, governmental corruption, and the ruthlessness of their (now former) president (Lubin, 2011). In a similar manner to Tunisian cables, the Egyptian cables also showed that the U.S., even though it supported the president of Egypt, was also helping elements of the opposition. The cables that WikiLeaks released were also interpreted by some as giving evidence that the U.S. government had been planning to topple the Egyptian president for the past three years (Slier, 2011).

WikiLeaks has surely played a role in the radical changes occurring in the Middle East and North Africa, even though that role may be quite different from what Julian Assange claims it to be. In helping to remove these dictators from power, WikiLeaks is also knocking power out of U.S. hands, since each one of these dictators has helped support the American Empire. Kibaki has sworn to deepen ties with the U.S., and in 2009 said, "The process [of Americanization] has begun and I am confident that we will have a new constitution in the coming months" (Kenyan News Archive, 2009). Tunisian dictator Ben Ali has been called a "model U.S. client", and it has been said that he "exemplified what the U.S. believed serves its interests: a blend of neoliberalism that is open to foreign investment, cooperation with American anti-terrorism by way of extreme rendition of suspects, and strict secularism that translates into the repression of political expression" (Falk, 2011, ¶ 19). Egyptian president Mubarak was arguably the most U.S. supported of the three, with his regime receiving $1-2 billion annually in U.S. military aid alone. In fact, U.S. Vice-President Joe Biden refused to call him a dictator, saying, "Mubarak has been an ally of ours in a number of things. And he's been very responsible on, relative to geopolitical interest in the region...I would not refer to him as a dictator" (Lehrer, 2011, 3:20).

Injured Diplomacy

At this point, the damage dealt to the U.S. empire is mostly speculative (some writers have even hinted that WikiLeaks was one of the reasons why the Liberal Party in Canada achieved so few votes during the 2011 election [McArthur, Curry, Freeze & Friesen, 2011]), but solid evidence of a hurt U.S. is beginning to emerge from the haze of theories.

The majority of the U.S. diplomatic cables WikiLeaks exposed in "Cable Gate" were benign or embarrassing, but a handful seriously damaged U.S. diplomatic prowess. Karl Eikenberry, the U.S. Ambassador to Afghanistan, wrote one of these damaging cables on 26 February 2010. In the cable he called the Afghan Finance Minister Omar Zakhilwal, who is internationally quite well respected, "an extremely weak man" (Associated Press, 2010, ¶ 2). This choice of language has hurt more than Zakhilwal's feelings, who said, "to be honest with you, I am extremely saddened" (Associated Press, 2010, ¶ 7): the crude language has seriously impaired the already strained U.S.-Afghani relations. Zakhilwal said that because of the release of this cable, "It certainly will not be business as usual" (Associated Press, 2010, ¶ 3).

A similar event occurred with the U.S. Ambassador to Mexico, Carlos Pascal. WikiLeaks released a trove of cables from the U.S. State department that were colossally offensive to Mexican President Felipe Calderon, many of which were signed by Pascal. President Calderon said that these cables have caused "severe damage" to the country's relation to the U.S., and furthermore stated that he was so offended by what Pascal wrote that he can no longer work with the American ambassador in Mexico.

By revealing U.S. ambassadors' true opinions of their colleagues, WikiLeaks has shown the true face of U.S. diplomacy to the world. Speculatively, with this disillusionment comes something more important, a reduction of U.S. soft power. I believe that the offence that these cables caused will reduce the amount of U.S. bargaining power and in turn hurt (and possibly reduce) the stature of the U.S. Empire.

An Opaque Iron Curtain

Since WikiLeaks has empowered so many people, and boosted worldwide transparency, and has siphoned power from the American Empire, it may seem that all this organization can do is good. This popular and optimistic view is not held by everyone. Internet theorist and computer scientist Jaron Lanier believes that if governments became as open as WikiLeaks hoped then, "we'd turn into a closed society...like North Korea, [where] everyday life is militarized" (Lanier, 2010, p. 2, ¶ 11). Lanier has many criticisms of WikiLeaks, believing that WikiLeaks' focus on creating governmental transparency is hugely problematic:

"The Wikileaks method punishes a nation—or any human undertaking—that falls short of absolute, total transparency, which is all human undertakings, but perversely rewards an absolute lack of transparency. Thus an iron-shut government doesn't have leaks to the site, but a mostly-open government does". (Lanier, 2010, p. 1, ¶ 11)

In this model, Lanier believes that WikiLeaks is actually facilitating opacity of governments by punishing all who create leaks. Conversely, totalitarian governments that do not produce leaks bear no punishment at all. There might be some evidence to support Lanier's theory. From the 1970s to 2008 there were only three cases where leakers were punished by the U.S. government for their actions (Benjamin, 2011). In Obama's few short years, there have been six. Historically, this is an unprecedented amount of attempted government censorship, from the president who promised "Transparency and the rule of law will be the touchstones of this presidency" (TPMTV, 2009, 1:52). Of course not all of these prosecutions are directly related to WikiLeaks. The U.S. government is also acting out against WikiLeaks in an indirect and borderline illegal way. In November, just after Bradley Manning was arrested for allegedly leaking cables to WikiLeaks, MIT researcher and pro WikiLeaks activist David House was one of the many to be harassed by the U.S. Homeland Security. While returning from a short trip to Mexico, Homeland Security, without a warrant or reasonable grounds, detained him in the airport and confiscated all of his belongings (which have still not been returned to him) (Garris, 2010). This sort of tactic was used by the government in response to other types of activists (Burke, 2010), but this is the first time that it is being applied, so liberally, to whistleblowers. Steven Aftergood of the Federation of American Scientists has noticed this trend, and commented, "They're going after [whistleblowers] at every opportunity and with unmatched vigor" (Gerstein, 2010, ¶ 5). Of course it would be unfair to blame WikiLeaks for the actions taken by the U.S. government, which would be like blaming Tunisian protesters for the regime repressing them.

Lanier might look at this and believe that his model works, but I do not believe this is the case. The prosecution and (quasi-legal) harassment of U.S. citizens is a step away from transparency, but it is a small one. Inspiring truly international press, informing the public, fueling revolutions, and removing power from the American Empire are all things that WikiLeaks can rightly take credit for. Even if the organization overstates its impact, I believe WikiLeaks has caused a positive shift towards global transparency. Julian Assange's creation has started worldwide trends that affect everything from the U.S. State Department, to the media, to fledging democracies across the globe. WikiLeaks is truly in the process of changing the world.

References

Anderson, C. (2010). Julian Assange: Why the World Needs WikiLeaks [Video]. *TED*.
http://www.ted.com/talks/julian_assange_why_the_world_needs_wikileaks.html

Associated Press. (2010). Afghan Minister: WikiLeaks Hurt U.S. Afghan Ties. *CBS News*, December 5.
http://www.cbsnews.com/stories/2010/12/04/world/main7118263.shtml

Bacevich, A. (2005). *The New American Militarism: How Americans are Seduced by War*. Oxford, UK: Oxford University Press.

Ballmer, S. (2005). Unlimited Potential Grant Announcement. *Microsoft Corporation*, February 17.
http://www.microsoft.com/presspass/exec/steve/2005/02-17aacis.mspx

BBC. (2011). Assange Claims WikiLeaks Boosted Mid East Uprisings. *BBC News*, March 16.
http://www.bbc.co.uk/news/education-12758380

Benjamin, M. (2011). WikiLeakers and Whistle-Blowers: Obama's Hard Line. *TIME*, March 11.
http://www.time.com/time/nation/article/0,8599,2058340,00.html

Berkeley Graduate School of Journalism. (2010). The State of Play: The Fourth Annual Reva and David Logan Investigative Reporting Symposium [Video]. Logan Symposium: The New Initiatives, April 18.
http://fora.tv/2010/04/18/Logan_Symposium_The_New_Initiatives

Burke, T. (2010). Homeland Security Harasses Haiti Activists. *Fight Back News*, January 21.
http://www.fightbacknews.org/es/node/1777

Davis, M. (2011). Dateline: Assange Speaks [Video]. *SBS*, February 13.
http://www.sbs.com.au/dateline/story/webextra/id/600911/n/Assange-Speaks

Davis, N. & Leigh, D. (2010). Afghanistan War Logs: Massive Leak of Secret Files Exposes Truth of Occupation. *The Guardian*, July 25.
http://www.guardian.co.uk/world/2010/jul/25/afghanistan-war-logs-military-leaks

Elections Meter. (2011). Afghanistan War Popularity, Elections Meter.
http://electionsmeter.com/polls/afghanistan-war

Falk, R. (2011). Ben Ali Tunisia was a model US client. *Al Jazeera English*, January 25.
http://english.aljazeera.net/indepth/opinion/2011/01/201112314530411972.html

Garris, E. (2010). Bradley Manning Support Activist Raided by FBI. *Anti War*, November 20.
http://www.antiwar.com/blog/2010/11/10/bradley-manning-support-activist-raided-by-fbi/

Gerstein, J. (2010) Justice Dept. Cracks Down on Leaks. *Politico*, May 25.

http://www.politico.com/news/stories/0510/37721.html

Gettleman, J. (2007). Tribal Rivalry Boils Over After Kenyan Election. *The New York Times*, December 30.

http://www.nytimes.com/2007/12/30/world/africa/30cnd-kenya.html?ex=1356757200&en=a20d98232560946f&ei=5124&partner=permalink&exprod=permalink

Kennedy, J.F. (1963). 90th Anniversary of Vanderbilt University (May 18, 1963). Miller Center, University of Virginia, Presidential Speech Archive.

http://millercenter.org/scripps/archive/speeches/detail/3373

Kenyan News Archive. (2009). We Will Deepen Relations with the U.S.A, President Kibaki says. *Kenya State House*, June 9.

http://www.statehousekenya.go.ke/news/june09/2009180601.htm

Kroft, S. (2011). Sixty Minutes Overtime Interview with Julian Assange [Video]. *CBS News*, January 30.

http://www.cbsnews.com/video/watch/?id=7300036n&tag=segementExtraScroller;housing

Lamo, A. (2010). WikiLeaks: No Wiki, Just Leaks. *The Wikipedian*, July 31.

http://thewikipedian.net/2010/07/31/wikileaks-no-wiki-just-leaks/

Lanier, J. (2010). The Hazards of nerd Supremacy: The Case of WikiLeaks. *The Atlantic*, December 20.

http://www.theatlantic.com/technology/archive/2010/12/the-hazards-of-nerd-supremacy-the-case-of-wikileaks/68217/1/

Lehrer, J. (2011). Exclusive: Mubarak Is Not a Dictator, but People Have a Right to Protest [Video]. *PBS Newshour*, January 27.

http://www.pbs.org/newshour/bb/politics/jan-june11/biden_01-27.html?print=

Lubin, G. (2011). WikiLeaks Spurs on Protests by Releasing New Egypt Corruption Cables. *Business Insider*, January 28.

http://www.businessinsider.com/wikileaks-egypt-brutality-2011-1

McArthur, G., Curry, B., Freeze, C., & Friesen, J. (2011). U.S. Ambassador Questioned Ignatieff's Leadership, WikiLeaks Cables Reveal. *The Globe and Mail*, April 29

http://www.theglobeandmail.com/news/politics/us-ambassador-questioned-ignatieffs-leadership-wikileaks-cables-reveal/article2002792/

Peter, A. (2010). *The Greatest Quotations of All-Time*. Bloomington: Xlibris Corporation.

Reuters. (1989). Tide of Democracy Sweeping Over China, East Europe, Reagan Says. *Los Angeles Times*, June 14.

http://articles.latimes.com/1989-06-14/news/mn-2060_1_chinese-students-east-europe-totalitarianism

Shirky, C. (2011). WikiLeaks Has Created a New Media Landscape. *The Guardian*, February 4.

http://www.guardian.co.uk/commentisfree/2011/feb/04/wikileaks-created-new-media-landscape

Slier, P. (2011). Egypt Torn in Clashes for 6th Day in a Row, Over 100 killed. *Russia Today*, January 29.

http://rt.com/news/egyptian-president-dismisses-cabinet/

TPMTV. (2009). Obama: Transparency Will Be Touchstone [Video]. *CNN*, January 21.

http://www.youtube.com/watch?v=72g7qmeP1dE

Vallance, C. (2010). WikiLeaks and Iceland MPs propose "Journalism Haven". *BBC News*, February 10.

http://news.bbc.co.uk/2/hi/technology/8504972.stm

White, G. (2011). This is the WikiLeak that Sparked the Tunisian Crisis. *Business Insider*, January 14.

http://www.businessinsider.com/tunisia-wikileaks-2011-1

WikiLeaks. (2010). Collateral Murder. *WikiLeaks*.

http://www.collateralmurder.com/

———. (2011). About WikiLeaks. *WikiLeaks*.

http://213.251.145.96/About.html

CHAPTER SEVEN

DEMONS, ANGELS, AND THE MESSIAH: THE TOP TEN MYTHS IN THE WAR AGAINST LIBYA

Maximilian C. Forte*

Since Colonel Gaddafi has lost his military hold in the war against NATO and the insurgents/rebels/new regime, numerous talking heads have taken to celebrating this war as a "success". They believe this is a "victory of the Libyan people" and that we should all be celebrating. Others proclaim victory for the "responsibility to protect," for "humanitarian interventionism," and condemn the "anti-imperialist left". Some of those who claim to be "revolutionaries," or believe they support the "Arab revolution," somehow find it possible to sideline NATO's role in the war, instead extolling the democratic virtues of the insurgents, glorifying their martyrdom, and magnifying their role until everything else is pushed from view. I wish to dissent from this circle of acclamation, and remind readers of the role of ideologically-motivated fabrications of "truth" that were used to justify, enable, enhance, and motivate the war against Libya—and to emphasize how damaging the practical effects of those myths have been to Libyans, and to all those who favoured peaceful, non-militarist solutions.

These top ten myths are some of the most repeated claims, by the insurgents, and/or by NATO, European leaders, the Obama administration, the mainstream media, and even the so-called "International Criminal Court"—the main actors speaking in the war against Libya. In turn, we look at some of the reasons why these claims are better seen as imperial folklore, as the myths that supported the broadest of all myths—that this war is a "humanitarian intervention," one designed to "protect civilians". Again, the importance of these myths lies in their wide reproduction, with little question, and to deadly effect. In addition, they threaten to severely distort the ideals of human rights and their future invocation, as well aiding in the continued militarization of Western culture and society.

The Mythology of Humanitarian Intervention

Humanitarian war—that such polar opposites should be paired points to the workings of myth. When I use the word "myth" in this report, it is not necessarily meant to neatly correspond with the ways that anthropologists commonly studied and analyzed myth—but it might correspond in some broad ways. Myth is often deployed here in a way that, superficially at least, is much closer to popular understanding of myth as deception, and as erroneous belief that can best be tested and shown to be false—which is in fact almost the exact opposite of what anthropologists understood by myth, where myth transcends reality versus fiction, and stresses belief and truth (see Cohen, 1969, p. 337). Myths, for many older anthropologists, tended to be constructs based on references to the sacred and ancient, and were heavily tied up with rites (see Raglan, 1955, p. 454). "The chief characteristics of myth are as follows," explains Cohen, from the anthropologist's perspective:

> "a myth is a narrative of events; the narrative has a sacred quality; the sacred communication is made in symbolic form; at least some of the events and objects which occur in the myth neither occur nor exist in the world other than that of myth itself; and the narrative refers in dramatic form to origins or transformation". (1969, p. 337)

Myth is something that was supposedly opposed to "history" (Raglan, 1955, p. 454). But then I have to wonder just how different the myths of history that follow in the sections below really are when compared to what past generations of anthropologists explained. It may be that what unites myth understood as deception, and myth understood as beyond the real, is mystification.

Malinowski (1926, p. 13) argued that "myth fulfills in primitive culture an indispensable function; it expresses, enhances and codifies belief; it safeguards and enforces morality; it vouches for the efficiency and contains practical rules for the guidance of man". Pragmatic sanction, tied up with ritual, instructing people on proper behaviour—these are some of the other qualities of myth distilled by Raglan (1955, p. 457). Legitimating social institutions and practices are key qualities of myth in some theories; in others they are akin to dreams; or, in Durkheimian theory, a mode of categorizing the world (Cohen, 1969, p. 338, 341, 343). In Levi-Strauss' framework, myths are ways of mediating oppositions and contradictions (Cohen, 1969, p. 346). Some interpretations of history can also be mythical (Cohen, 1969, p. 352), for being an arbitrary selection of facts, which also attribute realities to events that they do not possess.

I would argue that the mythology of humanitarian interventionism bears many of these diverse qualities of myth studied by anthropologists. The myths stem from liberal ideology, which since the end of the Cold War enjoyed global dominance, perhaps unrivalled until recently. Intervention is a way of making a global culture of compliance and conformity. The mythology speaks the language of civilization, and opposes it to savagery—it generates a mode of categorizing the world, a taxonomical ordering that benefits political order and hierarchy. The mythology promises salvation, brought by the self-appointed Western messiah.

> "Here we go again. The cheering crowds. The deposed dictator. The encomiums to freedom and liberty. The American military as savior". (Hedges, 2011, ¶ 1)

The myths fabricate a world where there are rightful actors, and those acted upon; there are those whose legitimacy is never in question, and those whose legitimacy is always contingent on the good wishes of those in the dominant part of the world system. There is a basic division between good and evil—this is where we need demons and angels—and the good among us must "do something," while the evil ones say we must do nothing. These myths prescribe pragmatic action and moral sanction (or economic sanctions). The actions prescribed by myth promise the creation of a new world order. Bombing fulfills the role of a rite of passage, carrying a society from crisis, through war, and then reintegration into the world system as a newly fashioned object—something that is reborn. "Arab Spring," which implies rebirth, can be made to fit this narrative construction, if bombing is what the interventionists see as necessary to make the flowers bloom. It makes sense, seen in this framework, that former U.S. Secretary of State Condoleeza Rice should refer to the 2006 Israeli bombing of Lebanon as "the growing—the birth pangs of a new Middle East" (Rice, 2006, ¶ 54). Likewise, with reference to Libya, a Reuters article celebrates "the birth pangs of the world's newest democracy" (Maclean, 2011, ¶ 9)—this peculiar phrasing of Rice's is further echoed by neoconservative scholar Fouad Ajami (2011, ¶ 14): "the birth pangs of the new Libya". Intervention myths are apparently the new creation myths of our time.

Figure 7.1

A Qatar Emiri Air Force Mirage 2000-5 fighter jet takes off on 25 March 2011, from Souda Bay, Greece, on an Odyssey Dawn mission against Libya. (Source: U.S. Department of Defense photo by Paul Farley, U.S. Navy. In the public domain, under Title 17, Chapter 1, Section 105 of the U.S. Code.)

1. Genocide

Just a few days after the street protests began, on 21 February the very quick to defect Libyan deputy Permanent Representative to the UN, Ibrahim Dabbashi, stated: "We are expecting a real genocide in Tripoli. The airplanes are still bringing mercenaries to the airports" (Taki, 2011, ¶ 3). This is excellent: a myth that is composed of myths. With that statement he linked three key myths together—the role of airports (hence the need for that gateway drug of military intervention: the no-fly zone), the role of "mercenaries" (meaning, simply, black people—see Forte, 2011), and the threat of "genocide" (geared toward the language of the UN's doctrine of the Responsibility to Protect). As ham-fisted and wholly unsubstantiated as the assertion was, he was clever in cobbling together three ugly myths, one of them grounded in racist discourse and practice that endures to the present, with newer atrocities reported against black Libyan and African migrants on a daily basis. He was not alone in making these assertions. Among others like him, Soliman Bouchuiguir, president of the Libyan League for Human Rights, told Reuters on 14 March that if Gaddafi's forces reached Benghazi, "there will be a real bloodbath, a massacre like we saw in Rwanda"

(Abbas, 2011, ¶ 9). That is not the only time we would be deliberately reminded of Rwanda. Here was Lt. Gen Roméo Dallaire, the much worshipped Canadian force commander of the U.N. peacekeeping mission for Rwanda in 1994, currently an appointed senator in the Canadian Parliament and co-director of the Will to Intervene project at Concordia University. Dallaire, in a precipitous sprint to judgment, not only made repeated references to Rwanda when trying to explain Libya, he spoke of Gaddafi as "employing genocidal threats to 'cleanse Libya house by house'" (Dallaire & Bernstein, 2011, ¶ 1). This is one instance where selective attention to Gaddafi's rhetorical excess was taken all too seriously, when on other occasions the powers that be are instead quick to dismiss it: U.S. State Department spokesman, Mark Toner waved away Gaddafi's alleged threats against Europe by saying that Gaddafi is "someone who's given to overblown rhetoric" (Sawer, 2011, ¶ 12). How very calm, by contrast, and how very convenient—because as early as 23 February President Obama declared that he had instructed his administration to come up with a "full range of options" to take against Gaddafi (Branigin, Sheridan, & Lynch, 2011, ¶ 1). And as early as 25 February, NATO Secretary General Anders Fogh Rasmussen ominously indicated the first signs of a NATO move on Libya: in one tweet, he stated, "I have called for an emergency meeting in the North Atlantic Council today to discuss Libya," and in another, "The situation in Libya is of great concern. NATO can act as an enabler and coordinator if and when member states will take action" (Rasmussen, 2011c, 2011d) and NATO would be "prepared for any eventuality" (2011e, ¶ 2). These statements came just a single day after Rasmussen said "NATO as such has no plans to intervene" (2011b, ¶ 3; see also 2011a), but with the added caution that while Rasmussen did "not consider the situation in Libya a direct threat to NATO or NATO Allies...of course, there may be negative repercussions" (2011b, ¶ 1).

Figure 7.2

NATO Secretary General, Anders Fogh Rasmussen, speaking to a meeting of NATO ministers of defence, 10 March 2011. (Source: NATO website, in the public domain.)

"Genocide," unlike the careless way it was used as seen above, instead has a well established international legal definition, as seen repeatedly in the UN's 1948 Convention on the Prevention and Punishment of the Crime of Genocide, where genocide involves the persecution of a "a national, ethnical, racial or religious group" (United Nations [UN], 1948a, Art. 2). Not all violence is "genocidal". Internecine violence is not genocide. Genocide is neither just "lots of violence" nor violence against undifferentiated civilians. What Dabbashi, Dallaire, and others failed to do was to identify the persecuted national, ethnic, racial or religious group, and how it differed in those terms from those allegedly committing the genocide. They really ought to know better (and they do), one as a UN ambassador and the other as a much exalted expert and lecturer on genocide. This suggests that myth-making was either deliberate, or founded on the kind prejudice that proves the power of myth over those speaking for the myth.

What foreign military intervention did do, however, was to enable the actual genocidal violence that has been routinely sidelined until only very recently: the horrific violence against African migrants and black Libyans, singled out solely on the basis of their skin colour. That has proceeded without impediment, without apology, and until recently, without much notice. Indeed, the media even collaborated (see Forte 2011), rapid in asserting without evidence that any captured or dead black man must be a "mercenary". This is the genocide that the white, Western world, and those who dominate the "conversation" about Libya, has missed (and not by accident).

2. Gaddafi is "bombing his own people"

We must remember that one of the initial reasons in rushing to impose a no-fly zone was to prevent Gaddafi from using his air force to bomb "his own people"—a distinct phrasing that echoes what was tried and tested in the demonization of Saddam Hussein in Iraq. On 21 February, when the first alarmist "warnings" about "genocide" were being made by the Libyan opposition, both Al Jazeera (2011a) and the BBC claimed that Gaddafi had deployed his air force against protesters—as the BBC "reported": "Witnesses say warplanes have fired on protesters in the city" (2011a, ¶ 3). Yet, on 1 March, in a Pentagon press conference, when asked: "Do you see any evidence that he [Gaddafi] actually has fired on his own people from the air? There were reports of it, but do you have independent confirmation? If so, to what extent?" U.S. Secretary of Defense Robert Gates replied, "We've seen the press reports, but we have no confirmation of that". Backing him up was Admiral Mullen: "That's correct. We've seen no confirmation whatsoever" (U.S. Department of Defense, 2011, ¶ 54-56).

In fact, claims that Gaddafi also used helicopters and anti-aircraft guns against unarmed protesters are totally unfounded, a pure fabrication based on fake claims (Cockburn, 2011a, ¶ 2-3, 18). This is important since it was Gaddafi's domination of Libyan air space that foreign interventionists wanted to nullify, and therefore myths of atrocities perpetrated from the air took on added value as providing an entry point for foreign military intervention that went far beyond any mandate to "protect civilians".

David Kirkpatrick of *The New York Times*, as early as 21 March confirmed that, "the rebels feel no loyalty to the truth in shaping their propaganda, claiming nonexistent battlefield victories, asserting they were still fighting in a key city days after it fell to Qaddafi forces, and making vastly inflated claims of his barbaric behavior" (Kirkpatrick, 2011a, ¶ 6). The "vastly inflated claims" are what became part of the imperial folklore surrounding events in Libya that suited Western intervention. Rarely did the Benghazi-based journalistic crowd question or contradict their hosts.

3. Save Benghazi

This chapter is being written as the Libyan opposition forces march on Sirte and Sabha, the two last remaining strongholds of the Gaddafi government, with ominous warnings to the population that they must surrender, or else. Apparently, Benghazi became somewhat of a "holy city" in the international discourse dominated by leaders of the European Union and NATO. Benghazi was the one city on earth that could not be touched. It was like sacred ground. Tripoli? Sirte? Sabha? Those can be sacrificed, as we all look on, without a hint of protest from any of the powers that be—this, even as we get the first reports of how the opposition and government forces have slaughtered people in Tripoli (BBC, 2011d, 2011e). Let's turn to the Benghazi myth.

"If we waited one more day," Barack Obama said in his 28 March address, "Benghazi, a city nearly the size of Charlotte, could suffer a massacre that would have reverberated across the region and stained the conscience of the world" (Obama, 2011, ¶ 11). In a joint letter, Obama with UK Prime Minister David Cameron and French President Nicolas Sarkozy asserted: "By responding immediately, our countries halted the advance of Gaddafi's forces. The bloodbath that he had promised to inflict on the citizens of the besieged city of Benghazi has been prevented. Tens of thousands of lives have been protected" (Obama, Cameron, & Sarkozy, 2011, ¶ 3). Not only did French jets bomb a retreating column (IOL, 2011, ¶ 24-25; also see Figure 7.3), what we saw was a very short column of about 14 tanks, 20 armoured personnel carriers, some trucks and ambulances, and that clearly could have neither destroyed nor occupied Benghazi (EuroNews, 2011; Reuters, 2011a).

Other than Gaddafi's "overblown rhetoric," which the U.S. was quick to dismiss when it suited its purposes, there is to date still no evidence furnished that shows Benghazi would have witnessed the loss of "tens of thousands" of lives as proclaimed by Obama, Cameron, and Sarkozy. This was best explained by Professor Alan J. Kuperman:

> "The best evidence that Khadafy did not plan genocide in Benghazi is that he did not perpetrate it in the other cities he had recaptured either fully or partially—including Zawiya, Misurata, and Ajdabiya, which together have a population greater than Benghazi....Khadafy's acts were a far cry from Rwanda, Darfur, Congo, Bosnia, and other killing fields....Despite ubiquitous cellphones equipped with cameras and video, there is no graphic evidence of deliberate massacre....Nor

did Khadafy ever threaten civilian massacre in Benghazi, as Obama alleged. The 'no mercy' warning, of March 17, targeted rebels only, as reported by The New York Times, which noted that Libya's leader promised amnesty for those 'who throw their weapons away'. Khadafy even offered the rebels an escape route and open border to Egypt, to avoid a fight 'to the bitter end'". (Kuperman, 2011, ¶ 6-8, 9)

Figure 7.3

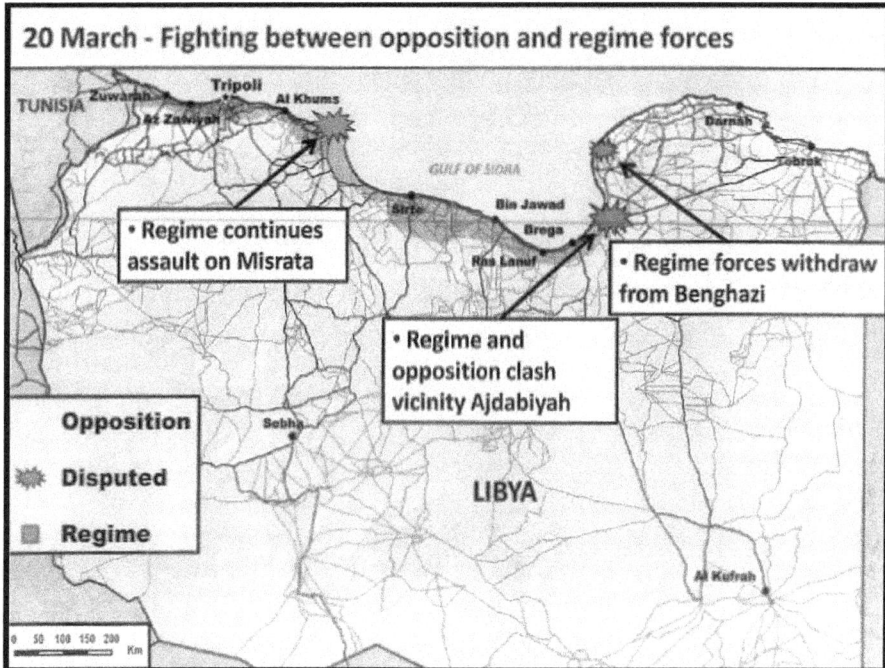

Map from a 20 March 2011 U.S. Department of Defense briefing showing fighting between rebel and government forces in Libya, and the withdrawal of Gaddafi's forces from Benghazi. (Source: U.S. Department of Defense. Provided by Wikimedia Commons under a Creative Commons License.)

In fact, during the uprising in Benghazi, Amnesty International found that no more than 110 people had been killed during the protests (including pro-government people), with another 59 to 64 in Baida (Cockburn, 2011a, ¶ 16)—still lower than the 232 persons killed in Cairo during the January protests in Egypt (Laub & Al-Shalchi, 2011, ¶ 7).

In a bitter irony, what evidence there is of massacres, committed by both sides, is now to be found in Tripoli in the period from 21 August, months after NATO imposed its "life-saving" military measures. Revenge killings are daily being reported with greater frequency, including the wholesale slaughter of black Libyans and African migrants by rebel forces (Cockburn, 2011b). Another sad irony: in Benghazi, which the insurgents have held for months now, not even that has prevented violence: revenge killings have been reported there too—more under Myth #6 below.

4. African Mercenaries

Libyans, who for the most part are Arabs, "generally do not consider themselves to be 'Africans' although the country is located on the African content"—even the "African" in Libya is "reserved for the country's minority of immigrants (many of whom are illegal) from sub-Saharan Africa" (Ghosh, 2011, ¶ 6). Before the revolt started, "there were an estimated 2.5 million migrant workers in Libya, perhaps the majority from black Africa," many remaining trapped in the country (Ghosh, 2011, ¶ 9; also see MacDougall, 2011). Cockburn (2011c, ¶ 6) offers the figure of an estimated one million illegal immigrants in Libya, which has a total population of six million and a workforce of 1.7 million. Given the long-standing racial discrimination against both black Libyans and migrant workers from Sub-Saharan Africa, coupled with Gaddafi's realignment of Libya with Africa, and his incorporation of black Libyans into this apparatus, it seems that the rebels have sought to exact revenge on Gaddafi and what are seen as his legions of African minions. It was reported that "some Libyan rebels seem to regard the war against Gadhafi as tantamount to a battle against black people" (Ghosh, 2011, ¶ 7). Amnesty International said that the new interim NTC regime has "made matters worse. They have ignited public anger by tapping into an existing xenophobia with very dire consequences for many guest workers" (Ghosh, 2011, ¶ 8).

Patrick Cockburn summarized the functional utility of the myth of the "African mercenary" and the context in which it arose:

> "Since February, the insurgents, often supported by foreign powers, claimed that the battle was between Gaddafi and his family on the one side and the Libyan people on the other. Their explanation for the large pro-Gaddafi forces was that they were all mercenaries, mostly from black Africa, whose only motive was money". (2011b, ¶ 11)

As he notes, black prisoners were put on display for the media (which is a violation of the Geneva Convention [UN, 1949, Art. 13]), but Amnesty International later found that all the prisoners had supposedly been released since none of them were fighters, but rather were undocumented workers from Mali, Chad, and west Africa. The myth was useful for the opposition to insist that this was a war between "Gaddafi and the Libyan people," as if he had no domestic support at all—an absolute and colossal fabrication such that one might question what opinion of the intellectual capacity of the audience is entertained by the producers of the myth. The myth is also useful for cementing the intended rupture between "the new Libya" and Pan-Africanism, realigning Libya with Europe and the "modern world" which some of the opposition so explicitly crave (more below).

The "African mercenary" myth, as put into deadly, racist practice, is a fact that paradoxically has been both documented and ignored. Elsewhere, about two months into the conflict, I provided an extensive review (Forte, 2011) of the role of the mainstream media, led by Al Jazeera (Mumisa, 2011), as well as the seeding of social media, in creating the African mercenary myth. Two of those responsible for giving credibility to falsehoods of African mercenaries landing in Libya were Mona El Tahawy, a frequent guest on CNN and Al Jazeera, and Dima Khatib, Al Jazeera's Latin America bureau chief (see Forte, 2011, ¶ 32, 32).

Among the departures from the norm of vilifying Sub-Saharan Africans and black Libyans that instead documented the abuse of these civilians, were the *Los Angeles Times* (Zucchino, 2011) and Human Rights Watch which found no evidence of any mercenaries at all in eastern Libya (Radio Netherlands Worldwide, 2011)—thus totally contradicting the claims presented as truth by Al Arabiya (2011) and *The Telegraph* (Ramdani, 2011), among others such as *TIME* (Perry, 2011) and *The Guardian* (Smith, 2011). In an extremely rare departure from the propaganda about the black mercenary threat which Al Jazeera and its journalists helped to actively disseminate, Al Jazeera produced a single report (AlJazeeraEnglish, 2011) focusing on the robbing, killing, and abduction of black residents in eastern Libya. That was soon after the protests began. Six months later, and now that the Associated Press (AP, 2011b), Channel 4 (2011), and others are noting the racism, Al Jazeera is trying to ambiguously show some interest (Al Jazeera, 2011c). Finally, there is also some increased recognition of these facts of media collaboration in the racist vilification of the insurgents' civilian victims (see Hart, 2011a; Kirkpatrick & Nordland, 2011).

Atrocities against black Libyans and African migrants in rebel-held areas were reported from early on, by a nominal number of reporters. Mary Fitzgerald, correspondent for *The Irish Times* in Tripoli, reported the following:

> "In eastern Libya earlier this year, I visited several facilities where suspected mercenaries were held. The men came from countries including Chad, Niger, Mali and Sudan. Some said they were innocent labourers who happened to be in the wrong place at the

wrong time; others claimed they had been tricked into fighting for Gadafy.

"In Baida, where a number of alleged mercenaries were said to have been hanged in front of the courthouse just after the town fell to the rebels, I saw graffiti which referred to abeed, the Arabic word for slaves, which can be used as a derogatory term for black people.

"In the rebel stronghold of Benghazi, I saw fighters bring a pick-up truck full of black men, their hands bound, to the seafront opposition headquarters one night. A hostile crowd milled around before they were driven off to a detention centre." (2011, ¶ 3-5)

A Turkish oilfield worker who fled Libya provided this horrifying account of what happened to his black African coworkers:

"We left behind our friends from Chad. We left behind their bodies. We had 70 or 80 people from Chad working for our company. They cut them dead with pruning shears and axes, attacking them, saying you're providing troops for Gadhafi. The Sudanese, the Chadians were massacred. We saw it ourselves". (Quoted in Quist-Arcton, 2011, ¶ 9)

The racist targeting and killing of black Libyans and Sub-Saharan Africans continues to the present. Patrick Cockburn (2011b) and Kim Sengupta (2011) speak of the recently discovered mass of "rotting bodies of 30 men, almost all black and many handcuffed, slaughtered as they lay on stretchers and even in an ambulance in central Tripoli". Even while showing us video of hundreds of bodies in the Abu Salim hospital, the BBC (2011c) dares not remark on the fact that most of those are clearly black people, and even wonders about who might have killed them. This is not a question for the anti-Gaddafi forces interviewed by Sengupta (2011, ¶ 4):

"'Come and see. These are blacks, Africans, hired by Gaddafi, mercenaries,' shouted Ahmed Bin Sabri, lifting the tent flap to show the body of one dead patient, his grey T-shirt stained dark red with blood, the saline pipe running into his arm black with flies. Why had an injured man receiving treatment been executed?"

Recent reports reveal the insurgents engaged in ethnic cleansing against black Libyans in Tawergha (Dagher, 2011; Black Star News [BSN], 2011), the insurgents calling themselves "the brigade for purging slaves, black skin," vowing that in the "new Libya" black people from Tawergha would be barred from health care and schooling in nearby Misrata (Kamara, 2011), from which black Libyans had already been expelled by the insurgents (Ford, 2011). Indeed, when Andrew Gilligan of *The Sunday Telegraph* visited Tawergha, what he found was an utterly depopulated town, none of the 10,000 inhabitants (mostly black Libyans) visible, and their homes thoroughly looted and vandalized. Abdul el-Mutalib Fatateth, the officer in charge of the rebel garrison in the town, told Gilligan: "We gave them thirty days to leave. We said if they didn't go, they would be conquered and imprisoned. Every single one of them has left, and we will never allow them to come back" (2011, ¶ 4). Gilligan reports, from his tour of the town, "we saw large numbers of houses, and virtually every shop, systematically vandalized, looted or set on fire" (2011, ¶ 8). He noted the "racist undercurrent" and confirmed that the Misrata brigade had indeed painted the slogan at the entrance to the town about purging black slaves and black skin (2011, ¶ 14-15). Another rebel commander declared to Gilligan: "Tawergha no longer exists" (2011, ¶ 29).

Currently, in detailing the racist-motivated abuses by the rebels, Human Rights Watch has reported:

> "Dark-skinned Libyans and sub-Saharan Africans face particular risks because rebel forces and other armed groups have often considered them pro-Gadhafi mercenaries from other African countries. We've seen violent attacks and killings of these people in areas where the National Transitional Council took control". (Abrahams, 2011, ¶ 6)

Amnesty International (2011b) also recently reported on the disproportionate detention of black Africans in rebel-controlled Az-Zawiya, as well as the targeting of unarmed, migrant farm workers. In Tripoli, fighters for the new regime have abducted thousands of black Libyans and Sub-Saharan African migrant workers, seizing anyone on the street who is black, and detaining them in a stadium on the accusation of being mercenaries (Hubbard, 2011; see also Gillis, 2011). Reports continue to mount as this is being written, with other human rights groups finding evidence of the insurgents targeting Sub-Saharan African migrant workers (AP, 2011b). As the chair of the African Union, Jean Ping, recently stated:

> "NTC seems to confuse black people with mercenaries. All blacks are mercenaries. If you do that, it means (that the) one-third of the population of Libya, which is black, is also mercenaries. They are killing people, normal workers, mistreating them". (AP, 2011b, ¶ 2)

An Amnesty International (2011c, ¶ 2, 3) delegation also personally witnessed the abuse and abduction of injured black patients in the Central Tripoli Hospital on 29 August by the rebels:

"An Amnesty delegation visiting the Central Tripoli Hospital on Monday witnessed three thuwwar revolutionaries (as the opposition fighters are commonly known) dragging a black patient from the western town of Tawargha from his bed and detaining him. The men were in civilian clothing.

"The thuwwar said the man would be taken to Misratah for questioning, arguing that interrogators in Tripoli 'let killers free'. Two other black Libyans receiving treatment in the hospital for gunshot wounds were warned by the anti-Gaddafi forces that 'their turn was coming'".

Amnesty International's Diana El Tahawy also observed that "What we are seeing in western Libya is a very similar pattern to what we have seen in Benghazi and Misurata after those cities fell to the rebels"—and Amnesty found around 300 people, mostly sub-Saharan Africans accused of being *mortazaga* (mercenaries), in detention in Zawiyah, 20 miles west of Tripoli, since the rebels took over (Van Langendonck, 2011, ¶ 8-9). (To read more, please consult the list of recent reports that I continue to compile.[1])

Reports from Amnesty International continued to come in during September (2011d, 2011e), with no apparent effort by the NTC to do anything to curb the violence against black Libyan and Sub-Saharan Africans. Indeed, the NTC Chairman himself, Mustafa Abdel Jalil, openly claimed in media interviews that "African mercenaries" were used by Gaddafi, and that when he served in Gaddafi's government, he witnessed that "40 per cent of criminals [in Libya] are Africans" (Amnesty International, 2011e, p. 83). Amnesty also reports that Jalil recently promised to close Libya to all Africans (2011e, p. 89). Another prominent NTC leader, Abd-al-Aziz al-Isawi, stated about the African presence in Libya, "They are a burden on health care, they spread disease, crime. They are illegal" (UN Watch, 2010, ¶ 9)—this was when he too served in Gaddafi's government, as Secretary of the General People's Committee of Libya (GPCO) for Economy, Trade and Investment. In his role in the NTC, responsible for foreign affairs, he has continued his demonizing diatribes against Africans, further perpetuating the mercenary myth and inventing crimes (see Smith, 2011, ¶ 4-5).

Amnesty recently reported that NTC rebels had also committed war crimes (2011d, ¶ 2). Amnesty reported that "when Al-Bayda, Benghazi, Derna, Misratah and other cities first fell under the control of the NTC in February, anti-Gaddafi forces carried out house raids, killings and other violent attacks against suspected mercenaries, either sub-Saharan Africans or black Libyans" (2011d, ¶ 9). The organization documented the abuse of prisoners, executions, and forced disappearances. It also noted that between a third and a half of all prisoners in Tripoli and al-Zawiya were either black Libyans or Sub-Saharan African migrant workers, further pointing out "that widespread rumours that al-Gaddafi forces used large numbers of sub-Saharan African mercenaries in February had been significantly exaggerated" (2011d, ¶ 12-13; also see 2011e, pp. 9, 12, 79, 81-86). Amnesty International also documented the lynching (its terms) of prisoners in rebel-controlled zones, with the emphasis of the brutality being against black Libyan soldiers (2011e, p. 71).

The "African mercenary" myth continues to be one of the most vicious of all the myths, and the most racist. Even in recent days, newspapers such as the *Boston Globe* uncritically and unquestioningly show photographs of black victims[2] or black detainees[3] with the immediate assertion that they must be mercenaries, despite the absence of any evidence. Instead we are usually provided with casual assertions that Gaddafi is "known to have" recruited Africans from other nations in the past, without even bothering to find out if those shown in the photos are black Libyans (Graff, 2011, ¶ 5). The lynching of both black Libyans and Sub-Saharan African migrant workers has been continuous, and has neither received any expression of even nominal concern by the U.S. and NATO members, nor has it aroused the interest of the so-called "International Criminal Court". There is as little chance of there being any justice for the victims as there is of anyone putting a stop to these heinous crimes that clearly constitute a case of ethnic cleansing. The media, only now, are becoming more conscious of the need to cover these crimes, having glossed them over for months. Amnesty International itself accuses EU member states and NATO for doing little more than "paying lip-service" to human rights, and cites this as an "international failure" to protect (2011e, p. 87).

None of this is meant to whitewash the sins of both the Gaddafi regime and ordinary Libyans against African migrants and refugees, both prior to the rebellion, and in the early stages of the rebellion (see Amnesty International, 2011e, pps. 79-80). Rather than the creation of a "new Libya" and the "protection of civilians" offering any kind of amelioration, however, the situation for black Libya has worsened even further, and for all of Gaddafi's supposed ambiguity, it is the rebels that black Libyans and African migrants fear, as the rebels are the only ones targeting them.

5. Viagra-fueled Mass Rape

The reported crimes and human rights violations of the Gaddafi regime are awful enough as they are that one has to wonder why anyone would need to invent stories, such as that of Gaddafi's troops, with erections powered by Viagra, going on a rape spree. Perhaps it was peddled because it's the kind of story that "captures the imagination of traumatized publics" (Murphy, 2011, ¶ 9). This story was taken so seriously that some people started writing to Pfizer to get it to stop selling Viagra to Libya, since its product was allegedly being used as a weapon of war. People who otherwise should know better, set out to deliberately misinform the international public.

The Viagra story was first disseminated by Al Jazeera (2011b), in collaboration with its rebel partners, favoured by the Qatari regime that funds Al Jazeera. It was then redistributed by almost all other major Western news media (see Iqbal, 2011b, for some examples).

Luis Moreno-Ocampo, Chief Prosecutor of the International Criminal Court, appeared before the world media to say that there was "information" (Bowcott, 2011, ¶ 11) that Gaddafi distributed Viagra (AFP, 2011b) to his troops in order "to enhance the possibility to rape" (Zirulnick, 2011, ¶ 2) and that Gaddafi ordered the rape of hundreds of women (BBC, 2011b, ¶ 1). Moreno-Ocampo insisted: "We are getting information that Qaddafi himself decided to rape" (Varner, 2011, ¶ 2) and that "we have information that there was a policy to rape in Libya those who were against the government" (Cockburn, 2011a, ¶ 4). He also exclaimed that Viagra is "like a machete," and that "Viagra is a tool of massive rape" (CNN, 2011a, ¶ 6).

In a startling declaration to the UN Security Council, U.S. Ambassador Susan Rice also asserted that Gaddafi was supplying his troops with Viagra to encourage mass rape (MacAskill, 2011). She offered no evidence whatsoever to back up her claim (NBC, 2011, ¶ 2). Indeed, U.S. military and intelligence sources flatly contradicted Rice, telling NBC News that "there is no evidence that Libyan military forces are being given Viagra and engaging in systematic rape against women in rebel areas" (2011, ¶ 1). Rice is a liberal interventionist who was one of those to persuade Obama to intervene in Libya. She utilized this myth because it helped her make the case at the UN that there was no "moral equivalence" between Gaddafi's human rights abuses and those of the insurgents.

U.S. Secretary of State Hillary Clinton also declared that "Gadhafi's security forces and other groups in the region are trying to divide the people by using violence against women and rape as tools of war, and the United States condemns this in the strongest possible terms" (Sidner, 2011, ¶ 34). She added that she was "deeply concerned" by these reports of "wide-scale rape" (Sidner, 2011, ¶ 32). (She has, thus far, said nothing at all about the rebels' racist lynchings.)

By 10 June, Cherif Bassiouni, who is leading a UN rights inquiry into the situation in Libya, suggested that the Viagra and mass rape claim was part of a "massive hysteria" (AFP, 2011c, ¶ 2). Indeed, both sides in the war have made the same allegations against each other. Bassiouni also told the press of a case of "a woman who claimed to have sent out 70,000 questionnaires and received 60,000 responses, of which 259 reported sexual abuse" (AFP, 2011c, ¶ 9). However, while his teams asked for those questionnaires, they never received them—"But she's going around the world telling everybody about it...so now she got that information to Ocampo and Ocampo is convinced that here we have a potential 259 women who have responded to the fact that they have been sexually abused," Bassiouni said (AFP, 2011c, ¶ 10-11). He also pointed out that it "did not appear to be credible that the woman was able to send out 70,000 questionnaires in March when the postal service was not functioning" (AFP, 2011c, ¶ 12). In fact, Bassiouni's team "uncovered only four alleged cases" of rape and sexual abuse: "Can we draw a conclusion that there is a systematic policy of rape? In my opinion we can't" (Nebehay, 2011, ¶ 13). In addition to the UN, Amnesty International's Donatella Rovera said in an interview with the French daily Libération, that Amnesty had,

> "not found cases of rape....Not only have we not met any victims, but we have not even met any persons who have met victims. As for the boxes of Viagra that Gaddafi is supposed to have had distributed, they were found intact near tanks that were completely burnt out" (Perrin, 2011, ¶ 5).

However, this did not stop some news manufacturers from trying to maintain the rape claims, in modified form. The BBC went on to add another layer just a few days after Bassiouni humiliated the ICC and the media: the BBC now claimed that rape victims in Libya faced "honour killings" (Harter, 2011). This is news to the few Libyans I know, who never heard of honour killings in their country. The scholarly literature turns up little or nothing on this phenomenon in Libya. The honour killings myth serves a useful purpose for keeping the mass rape claim on life support: it suggests that women would not come forward and give evidence, out of shame, and fear of being killed by their families. Also just a few days after Bassiouni spoke, Libyan insurgents, in collaboration with CNN (Sidner, 2011), made a last-ditch effort to save the rape allegations: they presented a cell phone with a rape video on it, claiming it belonged to a government soldier. The men shown in the video are in civilian clothes. There is no evidence of Viagra. There is no date on the video and we have no idea who recorded it or where. Those presenting the cell phone claimed that many other videos existed, but they were conveniently being destroyed to preserve the "honour" of the victims.

6. Responsibility to Protect (R2P)

Having asserted, wrongly as we saw, that Libya faced impending "geno-cide" at the hands of Gaddafi's forces, it became easier for Western powers to invoke the UN's 2005 doctrine of the Responsibility to Pro-tect (UN, 2005, ¶ 138-140; UN, 2009). Meanwhile, it is not at all clear that by the time the UN Security Council passed Resolution 1973 (UN, 2011) that the violence in Libya had even reached the levels seen in Egypt, Syria, and Yemen. The most common refrain used against critics of the selectivity of this supposed "humanitarian interventionism" is that *just because the West cannot intervene everywhere does not mean it should not intervene in Libya*. Perhaps, but that still does not explain why *Libya* was the chosen target. This is a critical point because some of the earli-est critiques of R2P voiced at the UN raised the issue of selectivity, of who gets to decide, and why some crises where civilians are targeted (say, Gaza) are essentially ignored, while others receive maximum con-cern, and whether R2P served as the new fig leaf for hegemonic geopoli-tics (see Chomsky, 2009).

The myth at work here is that foreign military intervention was guided by humanitarian concerns. To make the myth work, one has to willfully ignore at least three key realities. One is the new scramble for Africa, where Chinese interests are seen as competing with the West for access to resources and political influence, something that AFRICOM is meant to counter (Christensen & Swan, 2008; Moraff, 2007). Gaddafi challenged AFRICOM's intent to establish military bases in Africa (see Rice, 2008, ¶ 3). In a cable released by WikiLeaks, we learn that the Government of Libya, "On AFRICOM...has argued that any foreign military presence, regardless of mission, on the African continent would constitute unacceptable latter-day colonialism and would present an at-tractive target for al-Qaeda" (U.S. Embassy Tripoli, 2008, ¶ 8). AFRICOM has since become directly involved in the Libya intervention and specifically "Operation Odyssey Dawn" (AFRICOM, 2011; Skin-ner, 2011; Stevenson, 2011). Horace Campbell (2011b, ¶ 1) argued that "U.S. involvement in the Libyan bombing is being turned into a public relations ploy for AFRICOM" and an "opportunity to give AFRICOM credibility under the facade of the Libyan intervention" (2011a, ¶ 1). In addition, Gaddafi's power and influence on the continent had also been increasing, through aid, investment, and a range of projects designed to lessen African dependency on the West and to challenge Western multilateral institutions by building African unity (see Pougala, 2011)—rendering him a rival to U.S. interests. Secondly, one has to recognize not just the anxiety of Western oil interests (Mufson, 2011, ¶ 4) over Gaddafi's "resource nationalism" (threatening to take back what oil companies had gained) (U.S. Embassy Tripoli, 2007, ¶ 1), an anxiety now clearly manifest in the European corporate rush into Libya to scoop up the spoils of victory (see: Borger & Macalister, 2011; Shabi, 2011; Taylor, 2011; Tong, 2011; Upstream, 2011)—but one has to also ignore the apprehension over what Gaddafi was doing with those oil revenues in supporting greater African economic independence, and for historically backing national liberation movements that challenged Western hegemony. Thirdly, one has to acknowledge the fear in Washington that the U.S. was losing a grip on the course of the so-called "Arab revolution" (Schwartz, 2011; Terrill, 2011a, 2011b; Wall Street Journal, 2011).

The problem for those proposing that the war in Libya was a "humanitarian" intervention, is that in light of the above stacking of realities, humanitarianism seems to be the least of the concerns for U.S. and European strategists, which even go as far as touting a version of democracy that is structured in terms of greater stability and thus security against the "Islamist threat". Given the track record of U.S. (and NATO) interventions in Kosovo/Serbia, Afghanistan, and Iraq, in terms of a grisly pile of the interventionists' own human rights abuses, it seems less than convincing to argue that a commitment to "human rights" is the core motivation here.

If R2P is seen as founded on moral hypocrisy and contradiction—as seems to be the case—it will become much harder in the future to cry wolf again and expect to get a respectful hearing. This is especially the case since little in the way of diplomacy and peaceful negotiation preceded the military intervention, which is directly contrary to what R2P mandates. While Obama is accused by some of having been slow to react (Abrams, 2011; Aujla, 2011; Hamid, 2011; McCain & Graham, 2011; Memoli, 2011), this was if anything a rush to war, on a pace that by very far surpassed Bush's invasion of Iraq. Not only do we know from the African Union (AU, 2011, p. 1) about how its efforts to establish a peaceful transition were impeded, but Dennis Kucinich also reveals that he received reports that a peaceful settlement was at hand, only to be "scuttled by State Department officials" (Kucinich, 2011, ¶ 11). These are absolutely critical violations of the R2P doctrine, showing how those ideals could instead be used for a practice that involved a hasty march to war, and a war aimed at regime change (which is itself a violation of international law).

That R2P served as a justifying myth that often achieved the opposite of its stated aims, is no longer a surprise. I am not even speaking here of the role of Qatar and the United Arab Emirates in bombing Libya and aiding the insurgents—even as they backed Saudi military intervention to crush the pro-democracy protests in Bahrain, nor of the ugly pall cast on an intervention led by the likes of unchallenged abusers of human rights who have committed war crimes with impunity in Kosovo, Iraq, and Afghanistan. Instead I am taking a narrower approach—such as the documented cases where NATO not only willfully *failed to protect civilians* in Libya, but it even *deliberately and knowingly targeted them* in a manner that constitutes terrorism by most official definitions used by Western governments.

Concerning the targeting of civilians, NATO admitted to deliberately targeting Libya's state television, killing three civilian reporters, in a move condemned by international journalist federations as a direct violation of a 2006 Security Council resolution banning attacks on journalists (AP, 2011a; Kirkpatrick, 2011c; UN, 2006a). A U.S. Apache helicopter—in a repeat of the infamous killings shown in the Collateral Murder video—gunned down civilians in the central square of Zawiya, killing the brother of the information minister among others (CNN, 2011b, ¶ 2). Taking a fairly liberal notion of what constitutes "command and control facilities," NATO targeted a civilian residential space resulting in the deaths of some of Gaddafi's family members, including three grandchildren (Verkaik, 2011). As if to protect the myth of "protecting civilians" and the unconscionable contradiction of a "war for human rights," the major news media often kept silent about civilian deaths caused by NATO bombardments (FAIR, 2011). R2P was invisible when it came to protecting civilians targeted by NATO.

In terms of the failure to protect civilians, in a manner that is actually an international criminal offense, we have the numerous reports of NATO ships ignoring the distress calls of refugee boats in the Mediterranean that were fleeing Libya. In May, 61 African refugees died on a single vessel (Shenker, 2011), despite making contact with vessels belonging to NATO member states. In a repeat of the situation, dozens died in early August on another vessel (Simpson, 2011). In fact, on NATO's watch, at least 1,500 refugees fleeing Libya have died at sea since the war began (Schwarz, 2011, ¶ 1). They were mostly Sub-Saharan Africans, and they died in multiples of the death toll suffered by Benghazi during the protests. R2P was utterly absent for these people (see Amnesty International, 2011e, p. 87).

At the time of writing, that was still the case—even as Western nations and major NGOs, pledged new aid to the nascent NTC regime, lifted sanctions against Libya, and called for reconciliation and inclusion (CNN, 2011c; Lekic, 2011), African migrants in Libya and black Libyans continued to suffer increasing numbers of rape, robbery, murder, and arbitrary detention. Many remained stranded at the port in Tripoli, while others could not venture out to cross Tripoli to reach the port (France24, 2011; Hill, 2011). None of this even raised an eyebrow among the "Friends of Libya"—despite all of the late media attention that was finally bringing their plight to the fore.

NATO has developed a peculiar terminological twist for Libya, designed to absolve the rebels of any role in perpetrating crimes against civilians, and abdicating its so-called responsibility to protect. Throughout the war, spokespersons for NATO and for the U.S. and European governments consistently portrayed all of the actions of Gaddafi's forces as "threatening civilians," even when engaged in either defensive actions, or combat against armed opponents. For example, after the collapse of the Gaddafi regime, the NATO spokesperson, Roland Lavoie, "appeared to struggle to explain how NATO strikes were protecting civilians at this stage in the conflict. Asked about NATO's assertion that it hit 22 armed vehicles near Sirte, he was unable to say how the vehicles were threatening civilians, or whether they were in motion or parked" (Laub & Schemm, 2011, ¶ 8). Indeed, NATO can no longer explain its continued intervention and bombing, because according to its own statement of three objectives, issued on 14 April (see NATO, 2011a, ¶ 5), all of the conditions stipulated for ending NATO operations had been achieved. None of the Gaddafi strongholds are listed as areas of NATO concern, obviously, since he would pose no threat to his bastions of support. Nonetheless, NATO is bombing those cities.

As the tables were turned, and Sirte became the new Benghazi, R2P was not to be seen or heard—another clear case as to why the doctrine has been defeated by NATO. Had it really committed itself to simply protecting civilians, NATO would have bombed forces on both sides of the conflict, and the result would have been a political stalemate, which the interventionists were obviously not prepared to accept. This adventure was not about stopping the killing of civilians; it was instead about who would get to do so with impunity.

"We destroy in order to save," writes Hedges (2011, ¶ 4), and the bombing of Sirte "mocks the justification for intervention". Hedges adds: "Our intervention, as in Iraq and Afghanistan, has probably claimed more victims than those killed by the former regime. But this intervention, like the others, was never, despite all the high-blown rhetoric surrounding it, about protecting or saving Libyan lives. It was about the domination of oil fields by Western corporations".

By protecting the rebels, in the same breath as they spoke of protecting civilians, it is clear that NATO intended for us to see Gaddafi's armed opponents as mere civilians. Interestingly, in Afghanistan, where NATO and the U.S. fund, train, and arm the Karzai regime in attacking "his own people" (like they do in Pakistan), the armed opponents are consistently labeled "terrorists" or "insurgents"—and even if the majority of them are civilians who have never served in any official standing army, NATO does not call them civilians, as it does in Libya. They are insurgents in Afghanistan, and their deaths at the hands of NATO are listed separately from the tallies for civilian casualties. By some twist of logic, in Libya, armed fighters are treated by NATO as "civilians". In response to the announcement of the UN Security Council voting for military intervention, a volunteer translator for Western reporters in Tripoli made this key observation: "Civilians holding guns, and you want to protect them? It's a joke. We are the civilians. What about us?" (Richter, 2011, ¶ 10).

NATO has provided a shield for the insurgents in Libya to victimize unarmed civilians in areas they came to occupy. There was no hint of any "responsibility to protect" in these cases. NATO assisted the rebels in starving Tripoli of supplies, subjecting its civilian population to a siege that deprived them of water, food, medicine, and fuel (Kirkpatrick, 2011b). When Gaddafi was accused of doing this to Misrata, the international media were quick to cite this as a war crime (Stephen, 2011).

Save Misrata, kill Sirte—how the label humanitarian comes to be applied in such cases defies logic, and that is because myth is not designed to be tested by logic. Leaving aside the documented crimes by the insurgents against black Libyans and African migrant workers, the insurgents were also found by Human Rights Watch to have engaged in "looting, arson, and abuse of civilians in [four] recently captured towns in western Libya" (HRW, 2011). In Benghazi, which the insurgents have held for months now, revenge killings have been reported by *The New York Times* as late as this May (Fahim, 2011), and by Amnesty International (2011a) in late June, with the insurgents' National Transitional Council clearly faulted by Amnesty.

The responsibility to protect? It would now sound like something that has fallen into irreparable disrepute as a shabby fig-leaf for imperialism (Escobar, 2011a). With some 8,000 bombing raids, and over

30,000 bombs dropped, it's understandable then that some would declare, in disdain and disbelief: "One hell of a humanitarian intervention isn't it?" (Mountain, 2011, ¶ 1).

7. Gaddafi—the Demon

Figure 7.4

Col. Muammar Gaddafi, 30 November 2006. (Provided by Wikimedia Commons under a Creative Commons License, via Agência Brasil.)

Depending on one's perspective, either Gaddafi is a heroic revolutionary, and thus the demonization by the West is extreme, or Gaddafi is a really bad man, in which case the demonization is unnecessary and absurd. The myth here is that the history of Gaddafi's power was marked only by atrocity—he is thoroughly evil, without any redeeming qualities, and anyone accused of being a "Gaddafi supporter" should somehow feel more ashamed than those who, for example, openly support NATO with its own ample record of human rights violations. This is binary absolutism at its worst—virtually no one made allowance for the possibility that some might neither support Gaddafi, the insurgents, nor NATO. Everyone was to be forced into one of those camps, no exceptions allowed. What resulted was a phony debate, dominated by fanatics of one side or another. Missed in the discussion was recognition of the obvious: however much Gaddafi had been "in bed" with the West over the past decade, his forces were now fighting against a NATO in Libya.

The other result was the impoverishment of historical consciousness, and the degradation of more complex appreciations of the full breadth of the Gaddafi record. This would help explain why some would not rush to condemn and disown the man (and why we should not resort to crude and infantile caricaturing of their motivations). While even Glenn Greenwald feels the need to dutifully insert, "No decent human being would possibly harbor any sympathy for Gadaffi" (Greenwald, 2011, ¶ 7), I have known decent human beings in Nicaragua, Trinidad, Dominica, and among the Mohawks in Montreal who very much appreciate Gaddafi's support—not to mention his support for various national liberation movements, including the struggle against apartheid in South Africa. Some cannot reconcile what appear to be the many faces of Gaddafi, and demand that we choose one over the other as the real face. Some of these faces are seen by his domestic opponents, others are seen by recipients of his aid, and others were smiled at by the likes of Silvio Berlusconi,[4] Nicolas Sarkozy,[5] Condoleeza Rice,[6] Hillary Clinton (Figure 7.6), and Barack Obama (see Figure 7.5).[7] There are many faces, and they are all simultaneously real. Some refuse to "disown" Gaddafi, to "apologize" for his friendship towards them, no matter how distasteful, indecent, and embarrassing other "progressives" may find him. Indeed, Nelson Mandela's refusal to apologize for receiving support from Gaddafi in the fight against apartheid has been memorialized in song (Guanaguanare, 2010). Ironically, we support many dictators, with our very own tax dollars, and we routinely offer no apologies for this fact.

Figure 7.5

Muammar Gaddafi at G-8 summit in L'Aquila, Italy, 09 July 2009. Gaddafi is at far right. Centre-left: Nicolas Sarkozy and Barack Obama, and behind them is Stephen Harper. (Source: The Official White House Photostream, Pete Souza. Provided by Wikimedia Commons under a Creative Commons License.)

Speaking of the breadth of Gaddafi's record, that ought to resist simplistic, revisionist reduction, some might wish to note that even now, the U.S. State Department's webpage on Libya[8] still points to a Library of Congress Country Study[9] on Libya that features some of the Gaddafi government's many social welfare achievements over the years in the areas of medical care,[10] public housing,[11] and education.[12] In addition, Libyans have the highest literacy rate in Africa, according to the United Nations Development Programme and Libya is the only continental African nation to rank "high" in the UNDP's Human Development Index (see UNDP, 2009, p. 171). Even the BBC recognized these achievements:

> "Women in Libya are free to work and to dress as they like, subject to family constraints. Life expectancy is in the seventies. And per capita income—while not as high as could be expected given Libya's oil wealth and relatively small population of 6.5m—is estimated at $12,000 (£9,000), according to the World Bank. Illiteracy has been almost wiped out, as has homelessness—a chronic problem in the pre-

Gaddafi era, where corrugated iron shacks dotted many urban centres around the country". (Hussein, 2011, ¶ 3-4)

So if one supports health care, does that mean one supports dictatorship? And if "the dictator" funds public housing and subsidizes incomes, do we simply erase those facts from our memory? In addition, these facts point a number of suggestions: that Libyans did not rebel because of poverty, and there is as yet no respected theory of political revolution that explains how human rights abuses lead to revolt, leaving aside the fact that what "we" often consider to be "human rights" selectively leaves out the social and economic rights which Gaddafi had success in securing for most Libyans.

Figure 7.6

U.S. Secretary of State Hillary Clinton meets with Dr. Mutassim Qadhafi, National Security Adviser of Libya at the State Department, on 21 April 2009. (Source: http://video.state.gov/en/video/20345898001/meeting-with-national-security-adviser-of-libya/. In the public domain, under Title 17, Chapter 1, Section 105 of the U.S. Code.)

This speaks to a broader point: the fundamental narrowing and limiting of "human rights" to individual civil liberties alone, forgetting that the Universal Declaration of Human Rights included social and economic rights within the frame of human rights (UN, 1948b, Arts. 21-27), that is, basic and inalienable rights further codified in the UN's International Covenant on Economic, Social and Cultural Rights (UN, 1976). Unfortunately, in reflecting a Western liberal bias toward the individual, with an almost exclusive preference for civil liberties, Western human rights organizations and states that produce judgments of any given country's "human rights record" almost always exclude any discussion of social and economic rights. Fewer still seem to recognize that "foreign military intervention, aggression or occupation" is understood at the UN as a "violation of human rights, especially the right to self-determination" (UN, 2006b, p. 2). Even fewer interlocutors debating the war in Libya understood that regime change is itself a violation of international law. To call this a war for human rights is an atrocious reversal on many fronts.

Finally, what is also interesting to note is that despite Western leaders' recent courting of Gaddafi, the war has been used by some—notably the Obama administration and U.S. mainstream media—to walk back in time and re-create Libya under Gaddafi as a terrorist threat (often ignoring Gaddafi's own war against Al Qaeda, which precedes that of the U.S. by several years at least, and ignoring how the UK for a time backed Al Qaeda against Gaddafi—see Bright [2002]). One of the later myths to be added in the war against Libya involved a revival of the myth of Libya's role in the bombing of PanAm flight 103 over Lockerbie, Scotland, in 1988—with the renewed vilification of Abdelbasset Al-Megrahi, the man conveniently assumed to be the bomber (see Hart, 2011b; Herman, 2009).

8. Freedom Fighters—the Angels

The complement to the demonization of Gaddafi was the angelization of the "rebels". My aim here is not to counter the myth by way of inversion, and demonizing all of Gaddafi's opponents, who have many serious and legitimate grievances, and in large numbers have clearly had more than they can bear. There are many who should also answer for the war crimes they committed during this war, and for encouraging and conducting racist lynchings. I am especially interested in how "we," in the North Atlantic part of the equation, construct them in ways that suit our intervention. This is not to say that the "rebels" were not actively playing to the audience, by creating self-representations that appealed to Western liberals, and concomitantly echoing the so-called "universal values" of "freedom" and "democracy," even embracing the

Reaganesque eponym of "freedom fighters". They showed all the signs Western liberals and their pseudo-left partners wished to see, adorned with fancy websites and Facebook pages.13

One standard way that Western liberals constructed the rebels, repeated in different ways across a range of media and by U.S. government spokespersons, can be seen in this *New York Times'* depiction of the rebels as "secular-minded professionals—lawyers, academics, businesspeople—who talk about democracy, transparency, human rights and the rule of law" (Kirkpatrick, 2011a, ¶ 5). The listing of professions familiar to the American middle class which respects them, is meant to inspire a shared sense of identification between readers and the Libyan opposition, especially when we recall that it is on the Gaddafi side where the forces of darkness dwell: the main "professions" we find are torturer, terrorist, and African mercenary. Two authors, one who works for Democracy Now! and the other a student at New York University, imagined the insurgents as "reformist revolutionaries" (not understanding that the two terms represent very different political transformations), and instructed us on "the imperative for solidarity" with the insurgents, as if impatient with even what was then a debate that had only just begun (see Kamat & Shokr, 2011, p. 14). The same demand for solidarity, and support for a NATO military intervention as a means of *doing something*, using the world's leading counter-revolutionary militaries to supposedly support "fellow revolutionaries" was to be found echoed over and over, as if the immediate answer to the request for intervention must have been "yes". Furuhashi (2011, ¶ 9) put it well when he wrote,

> "What remains mysterious is why so many leftists, Arabs, and Iranians, secular or religious, reformist or revolutionary, in the West or in the axis of resistance fell deeply and blindly in love with the Libyan rebels. No matter how much they love the rebels (who remain 'revolutionaries' in the eyes of the slaves of love), there is no evidence whatsoever that the rebels love them or what they stand for".

This was indeed "fictive kinship" at its best, translated to the global arena, and at its best precisely for being so unquestioning, loading itself with the burden of obligation.

For many weeks it was almost impossible to get reporters embedded with the rebel National Transitional Council in Benghazi to even begin to provide a description of who constituted the anti-Gaddafi movement, if it was one organization or many groups, what their agendas were, and so forth. The subtle leitmotif in the reports was one that cast the rebellion as entirely spontaneous and indigenous—which may be true, in part, but it is also certainly an oversimplification. Among the reports that significantly complicated the picture were those that discussed the CIA ties to the insurgents (see: Fadel, 2011; Martin, 2011a, 2011b; and, WSWS, 2011). Others highlighted the role of the National Endowment for Democracy, the International Republican Institute, the National Democratic Institute, and USAID, which have been active in Libya since 2005 (Iqbal, 2011a). Then there were those that detailed the role of various expatriate groups (Scott, 2011); and, reports of the active role of "radical Islamist" militias (Jacinto, 2011) embedded within the overall insurgency, with some pointing to Al Qaeda connections (Escobar, 2011b).

Some feel a definite need for being on the side of "the good guys," especially as neither Iraq nor Afghanistan offer any such sense of righteous vindication. Americans want the world to see them as doing good, as being not only indispensable, but also irreproachable. They could wish for nothing better than being seen as atoning for their sins in Iraq and Afghanistan—it must come as surprising and welcome flattery to hear the following from the Islamic jihadist quarter of Libya: "They [the Americans] have started to redeem themselves for their past mistakes by helping us to preserve the blood of our children" (BBC, 2011f, ¶ 33). This is a special moment, where the bad guy can safely be the other once again. A world that is safe for America is a world that is unsafe for evil. War permits Americans "to wallow in unchecked self-exaltation. We are a nation that loves to love itself" (Hedges, 2011, ¶ 1). Marching band, Hellfire missiles, baton twirlers, Anderson Cooper, confetti, fragmentation bombs.

9. Victory for the Libyan People

To say that the current turn in Libya represents a victory by the Libyan people in charting their own destiny is, at best, an oversimplification that masks the range of interests involved since the beginning in shaping and determining the course of events on the ground, and that ignores the fact that for much of the war Gaddafi has been able to rely on a solid base of popular support. What is interesting to note is the strong, absolute, and early lead that European powers took in deciding on what would happen in Libya—and certainly without consulting the masses of protesters in Libya first.

As early as 25 February, a mere week after the start of the first street protests, Nicolas Sarkozy had already determined that Gaddafi "must go". By 28 February, David Cameron began working on a proposal for a no-fly zone—these statements and decisions were made without any attempt at dialogue and diplomacy.[14] By 30 March, *The New York Times* reported that for "several weeks" CIA operatives had been working inside Libya, which would mean they were there from mid-February, that is, when the protests began—they were then joined inside Libya by "dozens of British special forces and MI6 intelligence officers" (Mazetti & Schmitt, 2011, ¶ 2,3). The NYT also reported that "several weeks" before (again, around mid-February), President Obama "signed a secret finding authorizing the CIA to provide arms and other support to Libyan rebels" (Mazetti & Schmitt, 2011, ¶ 9), with that "other support" entailing a range of possible "covert actions" (Thomas, 2011, ¶ 4). USAID had already deployed a team to Libya by early March (DipNote, 2011; Lee, 2011). At the end of March, Obama publicly stated that the objective was to depose Gaddafi (Reuters, 2011b). In terribly suspicious wording, "a senior U.S. official said the administration had hoped that the Libyan uprising would evolve 'organically,' like those in Tunisia and Egypt, without need for foreign intervention" (Richter, 2011, ¶ 18)—which sounds like exactly the kind of statement one makes when something begins in a fashion that is not "organic" and when comparing events in Libya as marked by a potential legitimacy deficit and lack of critical mass when compared to those of Tunisia and Egypt. In one of many signals of overconfidence, showing just how weak the rebels were on the ground, on 14 March the NTC's Abdel Hafeez Goga asserted, "We are capable of controlling all of Libya, but only after the no-fly zone is imposed" (Bradley & Levinson, 2011, ¶ 25)—which is still not the case even six months later, not even after the fall of Tripoli.

In recent days it has also been revealed that what the rebel leadership swore it would oppose—"foreign boots on the ground" (AFP, 2011a)—is in fact a reality confirmed by NATO: "Special forces troops from Britain, France, Jordan and Qatar on the ground in Libya have stepped up operations in Tripoli and other cities in recent days to help rebel forces as they conducted their final advance on the Gadhafi regime" (Starr, 2011, ¶ 1). This, and other summaries,[15] are only scratching the surface of the range of external support provided to the rebels. The myth here is that of the nationalist, self-sufficient rebel, fueled entirely by popular support.

At the moment, war supporters are proclaiming the intervention a "success". It should be noted that there was another case where an air campaign, deployed to support local armed militia on the ground, aided by U.S. covert military operatives, also succeeded in deposing another regime, and even much more quickly. That case was Afghanistan. Success, anyone?

10. Defeat for "the Left"

As if reenacting the pattern of articles condemning "the left" that came out in the wake of the Iran election protests in 2009 (see as examples Hamid Dabashi [2009] and Slavoj Žižek [2009]), the war in Libya once again seemed to have presented an opportunity to target the left, as if this was topmost on the agenda—as if "the left" was the problem to be addressed. Here we see articles, in various states of intellectual and political disrepair, by Juan Cole (2011) (see some of the rebuttals: North [2011]; Van Aucken [2011a, 2011b]); Gilbert Achcar (2011a, and especially 2011b); Immanuel Wallerstein (2011); and, Helena Sheehan (2011) who seemingly arrived at some of her most critical conclusions at the airport at the end of her very first visit to Tripoli.

There seems to be some confusion over roles and identities. There is no homogeneous left or ideological agreement among anti-imperialists (which includes conservatives and libertarians, anarchists and Marxists). There is also confusion about anti-imperialism, notable when asking anti-imperialists to get on board with imperialism. Nor was the "anti-imperialist left" in any position to either do real harm on the ground, as is the case of the actual protagonists. There was little chance of the anti-interventionists influencing foreign policy, which took shape in Washington before any of the serious critiques against intervention were published. These points suggest that at least some of the critiques are moved by concerns that go beyond Libya, and that even have very little to do with Libya ultimately.

The most common accusation is that the anti-imperialist left is somehow coddling a dictator. This is supposed to embarrass the left—yet the accusers lack embarrassment of their own celebratory reaffirmation of their own political system as a beacon of democracy. The argument is that this anti-imperialist snuggling with dictators is based on a flawed analysis—in criticizing the position of Hugo Chávez, Wallerstein (2011, ¶ 4) says Chávez's analysis is deeply flawed, and offers this among the criticisms: "The second point missed by Hugo Chavez's analysis is that there is not going to be any significant military involvement of the western world in Libya" (yes, read it again). As a result, having dismissed the possibility of Western military intervention at various points in his commentary, Wallerstein concludes: "The issue therefore is not Western military intervention or not" (2011, ¶ 7). But since military intervention was indeed the issue and Chávez got his analysis right, then what would Wallerstein say now? We do not know, because like so many prominent commentators who openly supported military intervention (unlike Wallerstein, who never even saw it coming), he simply went silent on Libya from the moment NATO bombs began to drop.

Indeed, many of the counterarguments deployed against the anti-interventionist left echo or wholly reproduce the top myths that were dismantled above, that get their geopolitical analysis almost entirely wrong, and that pursue politics focused in part on personality and events of the day. This also shows us the deep poverty of politics premised primarily on very simple, absolute and one-sided ideas of "human rights" and "protection" (see Richard Falk's [2011] critique), and the success of the new military humanism in siphoning off the energies of the left.

Another common complaint was that those who opposed NATO's intervention privileged anti-imperialism over democracy and human rights. This is one very odd way of associating NATO with the struggle for freedom, democracy, and human rights—a record it simply cannot claim. Beyond what appeared to be naïve or willful acclamation for NATO, the basic problem overlooked by such formulations was that imperialism itself is dictatorship, at a global level, with a history of violating human rights of all kinds on a scale unmatched by any local dictator, and with a record of impeding, subverting, and overthrowing democratic regimes. One cannot have democracy and human rights under imperial domination, and that even extends to citizens of imperial states themselves to a growing degree. The idea that one can disentangle the human rights question from imperialism shows a remarkable non-understanding of what imperialism is and has been.

And a question persists: if those opposed to intervention were faulted for providing a moral shield for "dictatorship" (as if imperialism was not itself a global dictatorship), what about those humanitarians who have backed the rise of xenophobic and racist militants who by so many accounts engage in ethnic cleansing? Does it mean that the pro-interventionist crowd is racist? Do they even object to the racism? So far, I have heard only silence from those quarters—except for one recent, confused attempt to castigate the critics of racism in order to downplay racist violence and even NATO bombings (see Claiborne, 2011).

The agenda in brow-beating the anti-imperialist straw man masks an effort to curb dissent against an unnecessary war that has prolonged and widened human suffering; advanced the cause of war corporatists, transnational firms, and neoliberals; destroyed the legitimacy of multilateral institutions that were once openly committed to peace in international relations; violated international law and human rights; gave rise to racist violence; empowered the imperial state to justify its continued expansion; violated domestic laws; and reduced the discourse of humanitarianism to a clutch of simplistic slogans, reactionary impulses, and formulaic policies that privilege war as a first option.

No Lessons to Learn Here

The first question—*do you support foreign military intervention to support the rebels and save Libyans from being massacred*—was always a flawed question, not least of all for being the first question, and for being a question that demanded only one answer. The result is a paradox that can only satisfy the interventionists if they refuse to admit what transpired since 19 March: there was not going to be a massacre of the city of Benghazi, but certainly rebels would have been killed, just as certainly as they have been since NATO intervened, and just as certainly as many more Libyans have been killed than would have been the case had the revolt been defeated in March. Instead, the war drags on. If it was all meant to save lives, then it's mission unaccomplished: too many have died, and continue to die, to make the "saving lives" case. Some lives were saved, but at the expense of many more lost. Had Gaddafi won, would there not have been a wave of revenge killings? Possibly, but now that the rebels have won control we have seen a pattern of their own revenge killings. Is it less repugnant to the humanitarians? To endorse NATO as the protector of civilians defies logic—in terms of the "responsibility to protect" civilians, NATO has either refused to do what it was mandated to do, or it has ignored that responsibility, or it has done the opposite.

None of the options would have come without a death toll. Those who advocated foreign military intervention foresaw action that would last "days, not weeks" (Tapper, Khan, & Raddatz, 2011, ¶ 1), and later "weeks, not months" (Keaten, 2011, ¶ 1)—and could thus portray the intervention as one that would, at the end of the day, save more lives than would be otherwise be lost. Instead, the war did last months and is ongoing, and NATO's aerial campaign against Libya lasted twice as long as it did in Kosovo, even if fewer sorties have been flown over Libya to date (see NATO, 2011b, ¶ 5, and PBS, 2000, ¶ 16). It is quite obvious that intervention escalated the conflict and increased the number of those killed. It would thus appear that before 19 March, when NATO started its first bombing runs, there was a greater chance of Gaddafi achieving victory, and thus shortening the armed conflict, than there was of the anti-Gaddafi forces of achieving the same. Non-intervention would have been, ultimately, the most humanitarian of all of the options, *if saving lives had indeed been the primary concern.* However, it should be clear by now that saving lives was not the primary concern—it is which side ended the lives of the other side that counted the most. The decision to intervene was therefore political, not humanitarian.

At several points I have already mentioned how the Libya intervention is being touted as a success. Yes, the no-fly zone was successfully implemented: the Libyan Air Force never challenged it. However, the other measures of success failed to materialize with any of the promised rapidity, such as: 1) significant government or military defections to the side of the insurgents; 2) an ability by the insurgents to advance without the support of what is ultimately the world's biggest air force (an ability they never achieved); and, 3) popular uprisings against the Gaddafi regime across Libya (which did not happen). Instead, what we witnessed from early on included: 1) determined NATO involvement as a partisan to the conflict, with the rapid slippage from "protecting civilians," to clearly protecting the insurgents and aiding their military advance; 2) an expansion of the expected duration of the military intervention; 3) a rise in the fighting between government forces and insurgents; 4) mounting human rights abuses committed against African migrants and black Libyans; and, 5) an increased outflow of tens if not hundreds of thousands of refugees from towns at the centre of the increased hostilities. Ultimately, Gaddafi was indeed displaced from the centre of power—but the high cost should give rise to sobriety rather than cheerful pronouncements.

But what about "solidarity"? Indeed, what about the insurgents demanded solidarity? That they hate Gaddafi? Until February, most of the Western public barely mentioned Gaddafi in a conversation—he had become something like "old news," even the butt of late-night comedy shows, since the 1990s. Few protested when their political leaders, and defence contractors, were the ones to coddle Gaddafi. If hating Gaddafi was the signal virtue of the insurgents, then when did it start to matter to outsiders and why? After all, Gaddafi faced several armed uprisings and coup attempts before—and in the West there was no public clamor for his head when he crushed them.

Solidarity is exactly the sort of thing that you never demand, and never demand as an automatic response. Real solidarity is built with communication, exchange, reciprocity, mutual knowledge and trust. Those demanding solidarity could not even tell us who these rebels were, as they did not know themselves. Solidarity comes from earnest and transparent political practice, and not from taking morality hostage and engaging in emotional blackmail, while condemning those who would raise questions. It is only when you wish to produce hysteria that you resort to histrionics.

The first steps should have been to develop a peaceful transition, of the kind that the African Union advocated, for which it was never given a real chance. The friends of both sides of the conflict should have exercised pressure on their respective sides to back away from escalating violence—that also did not happen, indeed, the opposite occurred. The result is the destruction of a nation, and prospects for years of bloodshed. Whatever we may call all of this, "humanitarian" and "solidarity" do not recommend themselves as credible options. A success? Some are sufficiently satisfied with the results of the day now that historical amnesia appears to have set in, one that extends to forgetting the present: in one widely circulated essay, written during this period, Goldstein (2011) is convinced that wars are shorter, less bloody, and generally more humane, so much so that world peace may be at hand. Yet, examine his article for any references to Iraq. It's as if that savage slaughter had never taken place: no invasion and occupation, no sanctions before that, no previous war, no mass displacement and mass death that has literally decimated the population of Iraq. It is as if Libya had washed us of our sins, and now the skies look clearer, and the air seems cleaner. Move along, no atrocity to see here (Hart, 2011c).

We have been through this before. The impetus and the outcomes are so close to Kosovo that this looks like repetition. The parallels with Afghanistan are numerous. And once again, the "no-fly zone" has been robbed of its magic, if one examines the results with care. The things we thought we knew about Libya (Keating, 2011) were fantasy, replacing the things we ought to know about intervention, but instead mystified and misinformed the public through myth. It is fine for Chris Hedges (2011, ¶ 1) to now say, "You would think we would have learned in Afghanistan or Iraq. But I guess not," when, as he says, "stopping Gadhafi forces from entering Benghazi six months ago," is something "which I supported" (2011, ¶ 3). If he had indeed understood and learned, he would not have let himself be drawn in by the myths, to endorse the gateway intervention for NATO domination. Will we learn? Did we, last time?

Notes

* This article is a revised, expanded, and updated version of the article previously published as "The Top Ten Myths in the War against Libya," in *CounterPunch* (August 31, 2011): http://www.counterpunch.org/2011/08/31/the-top-ten-myths-in-the-war-against-libya/

1 http://www.diigo.com/user/openanthropology/Africans%2C%20racism

2 http://www.boston.com/bigpicture/2011/08/
libya_the_fight_continues.html#photo6

3 http://www.boston.com/bigpicture/2011/08/
libya_the_fight_continues.html#photo12

4 http://twitpic.com/6fv8vs

5 http://twitpic.com/6fv95a

6 http://twitpic.com/6dba10

7 http://twitpic.com/6aixl6

8 http://www.state.gov/p/nea/ci/ly/

9 http://lcweb2.loc.gov/frd/cs/lytoc.html

10 http://lcweb2.loc.gov/cgi-bin/query/r?frd/cstdy:@field(DOCID+ly0068)

11 http://lcweb2.loc.gov/cgi-bin/query/r?frd/cstdy:@field(DOCID+ly0069)

12 http://lcweb2.loc.gov/cgi-bin/query/r?frd/cstdy:@field(DOCID+ly0070)

13 The National Transitional Council of Libya, on Facebook:
http://www.facebook.com/libya.ntc and their English website at
http://www.ntclibya.com/Default.aspx?SID=1&ParentID=0&LangID=1

14 http://en.wikipedia.org/wiki/2011_military_intervention_in_Libya#Chronology

15 http://en.wikipedia.org/wiki/Timeline_of_the_2011_Libyan_civil_
war#Intelligence_operations_in_Libya

References

Abbas, M. (2011). Libya Jets Bomb Rebels, French Press For No-Fly Zone. *Reuters*,
March 14.
http://mobile.reuters.com/article/topNews/idUSTRE7270JP20110314?i=1&irp
c=932

Abrahams, F. (2011). A New Libya Must Honor Human Rights. *CNN*, August 24.
http://edition.cnn.com/2011/OPINION/08/23/abrahams.human.rights.libya/i
ndex.html?iref=allsearch

Abrams, E. (2011). Qaddafi's Fall: No, Obama Was Not Right. *National Review Online*, August 22.

http://www.nationalreview.com/corner/275207/qaddafi-s-fall-no-obama-was-not-right-elliott-abrams

Achcar, G. (2011a). Libyan Developments. *Znet*, March 19.

http://www.zcommunications.org/libyan-developments-by-gilbert-achcar

——— . (2011b). Libya: A Legitimate and Necessary Debate from an Anti-Imperialist Perspective. *Znet*, March 25.

http://www.zcommunications.org/libya-a-legitimate-and-necessary-debate-from-an-anti-imperialist-perspective-by-gilbert-achcar

AFP. (2011a). Europe's Invisible "Boots on the Ground" in Rebel Libya. *The Gazette*, May 8.

http://www.montrealgazette.com/news/Europe+invisible+boots+ground+rebel+Libya/4747380/story.html

——— . (2011b). Kadhafi 'Ordered Mass Rapes' in Libya: ICC. *Agence France Presse*, June 8.

http://www.google.com/hostednews/afp/article/ALeqM5iDLbC17PLbfZIhQo1y4gA-diS_0Q?docId=CNG.bc5038d72752104d215a218741da85d3.3e1

——— . (2011c). Libya Rape Claims 'Hysteria'—Investigator. *Herald Sun*, June 10.

http://www.heraldsun.com.au/news/breaking-news/libya-rape-claims-hysteria-investigator/story-e6frf7jx-1226072781882

African Union (AU). (2011). Press Statement of the 268th Meeting of the Peace and Security Council, PSC/PR/BR.1(CCLXVIII), March 23. Addis Ababa: Peace and Security Council, African Union.

http://www.au.int/en/sites/default/files/268th_FINAL_Press_Statement_-_Libya__EN.pdf

AFRICOM. (2011). AFRICOM Supports U.S. and International Response to Libya Crisis. *U.S. Africa Command*, March 5.

http://www.africom.mil/getArticle.asp?art=6134&

Ajami, F. (2011). From Baghdad to Tripoli. *The Wall Street Journal*, August 31.

http://online.wsj.com/article/SB10001424053111904332804576538583073952182.html?mod=googlenews_wsj

Al Arabiya. (2011). Gaddafi Recruits "African mercenaries" to Quell Protests. *Al Arabiya* (English), February 19.

http://www.alarabiya.net/articles/2011/02/19/138351.html

Al Jazeera. (2011a). Libya Protests Spread and Intensify: Diplomats Resign and Air Force Officers Defect as Gaddafi Government Resorts to Shooting and Bombing to Crush Uprising. *Al Jazeera English*, February 21.

http://english.aljazeera.net/news/africa/2011/02/2011221133557377576.html

——— . (2011b). Rape Used "As a Weapon" in Libya. *Al Jazeera English*, March 28.

http://english.aljazeera.net/video/africa/2011/03/201132845516144204.html

——— . (2011c). Foreign Migrants at Risk Libya. *Al Jazeera English*, August 29.

http://english.aljazeera.net/video/africa/2011/08/201182918514691615.html

AlJazeeraEnglish. (2011). Black Africans in Libya Live in Fear [Video]. March 1.

http://www.youtube.com/watch?v=YNA8z5G-Xmk

Amnesty International. (2011a). Document–African Union Must Prioritize the Protection of Civilians in Conflict Situations. *Amnesty International*, June 23. http://www.amnesty.org/en/library/asset/IOR63/002/2011/en/78a59751-37e4-4ed1-b32d-8743a8a21e44/ior630022011en.html

———. (2011b). Both Sides in Libya Conflict Must Protect Detainees from Torture. *Amnesty International*, August 25. http://www.amnesty.org/en/news-and-updates/both-sides-libya-conflict-must-protect-detainees-torture-2011-08-25

———. (2011c). Libya: Fears for Detainees Held by Anti-Gaddafi Forces. *Amnesty International*, August 31. http://amnesty.org.uk/news_details.asp?NewsID=19660

———. (2011d). Libya: NTC Must Take Control to Prevent Spiral of Abuses. *Amnesty International*, September 12. http://www.amnesty.org/en/news-and-updates/report/libya-ntc-must-take-control-prevent-spiral-abuses-2011-09-12

———. (2011e). *The Battle for Libya: Killings, Disappearances and Torture*. London: Amnesty International Ltd. http://www.amnesty.org/en/library/asset/MDE19/025/2011/en/8f2e1c49-8f43-46d3-917d-383c17d36377/mde190252011en.pdf

Associated Press (AP). (2011a). Media Group Urges UN Probe of Strike on Libya TV. *Associated Press*, August 5. http://news.yahoo.com/media-group-urges-un-probe-strike-libya-tv-172337644.html

———. (2011b). AU: Libya Rebels Killing Black Workers. *CBS News*, August 29. http://www.cbsnews.com/stories/2011/08/29/501364/main20099014.shtml

Aujla, S. (2011). McCain: Obama Should Have Acted Faster. *Politico*, March 20. http://www.politico.com/blogs/politicolive/0311/McCain_Obama_should_have_acted_faster_in_Libya.html

BBC. (2011a). Libya Protests: Tripoli Hit by Renewed Clashes. *British Broadcasting Corporation*, February 21. http://www.bbc.co.uk/news/world-africa-12531637

———. (2011b). Libya: Gaddafi Investigated Over Use of Rape as Weapon. *British Broadcasting Corporation*, June 8. http://www.bbc.co.uk/news/world-africa-13705854

———. (2011c). Libya: Hundreds of Bodies Found at Tripoli Hospital. *British Broadcasting Corporation*, August 26. http://www.bbc.co.uk/news/world-africa-14689451

———. (2011d). Libyan Capital Tripoli Faces Water, Power Crisis. *British Broadcasting Corporation*, August 27. http://www.bbc.co.uk/news/world-africa-14691061

———. (2011e). Libya: 'Mass killing' sites in Tripoli. *British Broadcasting Corporation*, August 31. http://www.bbc.co.uk/news/world-africa-14729083

———. (2011f). Fears over Islamists within Libyan Rebel Ranks. *British Broadcasting Corporation*, August 31.

http://www.bbc.co.uk/news/world-africa-14728565

Black Star News (BSN). (2011). Editorial: Ethnic Cleansing of Black Libyans. *Black Star News*, June 21.

http://www.blackstarnews.com/news/135/ARTICLE/7478/2011-06-21.html

Borger, J., & Macalister, T. (2011). The Race is on for Libya's Oil, with Britain and France Both Staking a Claim. *The Guardian*, September 1.

http://www.guardian.co.uk/world/2011/sep/01/libya-oil?CMP=twt_gu

Bowcott, O. (2011). Libya Mass Rape Claims: Using Viagra Would Be a Horrific First. *The Guardian*, June 9.

http://www.guardian.co.uk/world/2011/jun/09/libya-mass-rape-viagra-claim

Bradley, M., & Levinson, C. (2011). Arab League Urges Libya "No-Fly" Zone. *Wall Street Journal*, March 14.

http://webcache.googleusercontent.com/search?q=cache:y0FIrMHN_EgJ:online.wsj.com/article/SB10001424052748704838804576196681609529882.html+Arab+League+Urges+Libya+'No-Fly'+Zone&cd=1&hl=en&ct=clnk&gl=ca&source=www.google.ca

Branigin, W., Sheridan, M.B., & Lynch, C. (2011). Obama Condemns Violence in Libya, Asks For "Full Range of Options". *The Washington Post*, February 23.

http://www.washingtonpost.com/wp-dyn/content/article/2011/02/22/AR2011022206935.html

Bright, M. (2002). MI6 "Halted Bid to Arrest Bin Laden". *The Observer*, November 10.

http://www.guardian.co.uk/politics/2002/nov/10/uk.davidshayler

Campbell, H. (2011a). AFRICOM as Libya Bombing Motive. *Institute for Public Accuracy*, March 24.

http://www.commondreams.org/newswire/2011/03/24-5

————— . (2011b). Libya: U.S. Military and Africom—Between the Rocks and the Crusaders. *Pambazuka News*, March 31.

http://allafrica.com/stories/printable/201104020071.html

Channel 4. (2011). Evidence of Libya Massacre as Remains of 50 People Found. *Channel 4 News*, August 28.

http://www.channel4.com/news/evidence-of-libya-massacre-as-remains-of-50-people-found

Chomsky, N. (2009). Statement by Professor Noam Chomsky to the United Nations General Assembly: Thematic Dialogue on the Responsibility to Protect. New York: United Nations.

http://www.un.org/ga/president/63/interactive/protect/noam.pdf

Christensen, T.J., & Swan, J. (2008). China in Africa: Implications for U.S. Policy. *U.S. Africa Command*, June 5.

http://www.africom.mil/getArticle.asp?art=1786

Claiborne, C. (2011). Racism in Libya. *Daily Kos*, September 12.

http://www.dailykos.com/story/2011/09/12/1015087/-Racism-in-Libya?via=blog_511082

CNN. (2011a). ICC to Investigate Reports of Viagra-Fueled Gang-Rapes in Libya. *CNN*, May 17.

http://articles.cnn.com/2011-05-17/world/libya.rapes.icc_1_rapes-viagra-pills-libyan-leader-moammar-gadhafi?_s=PM:WORLD

——— . (2011b). Brother of Libya's Information Minister Reported Killed In NATO Strike. *CNN*, August 19.

http://edition.cnn.com/2011/WORLD/africa/08/18/libya.war/

——— . (2011c). "Friends of Libya" Converge on Paris. *CNN*, September 1.

http://edition.cnn.com/2011/WORLD/africa/08/31/libya.paris.meeting/

Cockburn, P. (2011a). Amnesty Questions Claim that Gaddafi Ordered Rape as Weapon of War. *The Independent*, June 24.

http://www.independent.co.uk/news/world/africa/amnesty-questions-claim-that-gaddafi-ordered-rape-as-weapon-of-war-2302037.html

——— . (2011b). Rebels Wreak Revenge on Dictator's Men. *The Independent*, August 28.

http://www.independent.co.uk/news/world/africa/rebels-wreak-revenge-on-dictators-men-2345261.html

——— . (2011c). The New Libya: Better Not Be Black. *CounterPunch*, August 30.

http://www.counterpunch.org/2011/08/30/the-new-libya/

Cohen, P.S. (1969). Theories of Myth. *MAN* (new series), 4(3), 337-353.

Cole, J. (2011). An Open Letter to the Left on Libya. *Informed Comment*, March 27.

http://www.juancole.com/2011/03/an-open-letter-to-the-left-on-libya.html

Dabashi, H. (2009). Left is Wrong on Iran. *Al-Ahram Weekly*, 956, July 16-22.

http://weekly.ahram.org.eg/2009/956/op5.htm

Dagher, S. (2011). Libya City Torn by Tribal Feud. *Wall Street Journal*, June 21.

http://online.wsj.com/article/SB10001424052702304887904576395143328336026.html

Dallaire, R., & Bernstein, J. (2011). Does the World Belong in Libya's War? Yes. Now Let's Hope It's Not Too Late. *Foreign Policy*, March 18.

http://www.foreignpolicy.com/articles/2011/03/18/does_the_world_belong_in_libyas_war?page=0,1

DipNote. (2011). U.S. Announces Additional Humanitarian Assistance in Response to Violence in Libya. *DipNote: U.S. Department of State Official Blog*, March 10.

http://blogs.state.gov/index.php/site/entry/additional_assistance_libya

Escobar, P. (2011a). R2P is Now Right 2 Plunder. *Asia Times Online*, August 27.

http://www.atimes.com/atimes/middle_east/mh27ak03.html

——— . (2011b). How al-Qaeda Got to Rule in Tripoli. *Asia Times Online*, August 30.

http://www.atimes.com/atimes/Middle_East/MH30Ak01.html

EuroNews. (2011). Libya Highway of Death 19 March 2011 [Video]. March 20.

http://www.youtube.com/watch?v=R-aDxW57IuQ

Fadel, L. (2011). Rebel Military Commander Wants to Be America's Man on the Ground in Libya. *The Washington Post*, April 12.

http://www.washingtonpost.com/world/rebel-military-commander-wants-to-be-americas-man-on-the-ground-in-libya/2011/04/12/AFF4g3SD_story.html

Fahim, K. (2011). Killings and Rumors Unsettle a Libyan City. *The New York Times*, May 10.

http://www.nytimes.com/2011/05/11/world/africa/11benghazi.html?pagewanted=all

FAIR. (2011). Media Advisory: Libyan Deaths, Media Silence. *FAIR–Fairness and Accuracy in Reporting*, August 18.

http://www.fair.org/index.php?page=4379

Falk, R. (2011). Can Humanitarian Intervention Ever be Humanitarian? *MWCNEWS*, August 4.

http://mwcnews.net/focus/editorial/12577-humanitarian-intervention.html?tmpl=component&print=1&layout=default&page=

Fitzgerald, M. (2011). "We are afraid . . . people might think we are mercenaries". *The Irish Times*, August 31.

http://www.irishtimes.com/newspaper/world/2011/0831/1224303238641.html

Ford, G. (2011). Black Libya City Said to Fall to Rebel Siege. *Black Agenda Report*, August 17.

http://blackagendareport.com/content/black-libya-city-said-fall-rebel-siege

Forte, M.C. (2011). The War in Libya: Race, "Humanitarianism," and the Media. *MRzine (Monthly Review Magazine)*, April 20.

http://mrzine.monthlyreview.org/2011/forte200411.html

France24. (2011). Ivorian Immigrant in Tripoli: "We blacks are brutalized because of our skin colour". *France24*, August 31.

http://observers.france24.com/content/20110831-libya-tripoli-ivorian-african-immigrant-blacks-brutalized-because-skin-colour-color

Furuhashi, Y. (2011). Loving the Libyan Rebels. *MRzine (Monthly Review Magazine)*, March 27.

http://mrzine.monthlyreview.org/2011/furuhashi270311.html

Ghosh, P.R. (2011). Libyan Revolt Unmasks Lethal Racism against Black Africans. *International Business Times*, August 31.

http://www.ibtimes.com/articles/206506/20110831/libya-revolt-rebels-black-african-migrants-racism-murder-gadhafi.htm

Gilligan, A. (2011). Gaddafi's Ghost Town After the Loyalists Retreat. *The Sunday Telegraph*, September 11.

http://www.telegraph.co.uk/news/worldnews/africaandindianocean/libya/8754375/Gaddafis-ghost-town-after-the-loyalists-retreat.html

Gillis, C.M. (2011). Mistaken for Mercenaries, Africans are Trapped in Libya. *Christian Science Monitor*, March 3.

http://www.csmonitor.com/World/Middle-East/2011/0303/Mistaken-for-mercenaries-Africans-are-trapped-in-Libya

Goldstein, J.S. (2011). Think Again: War. *Foreign Policy*, September-October.

http://www.foreignpolicy.com/articles/2011/08/15/think_again_war?page=full

Graff, P. (2011). Thirty Gaddafi Fighters Found Dead at Tripoli Camp. *Reuters*, August 25.

http://old.news.yahoo.com/s/nm/20110825/wl_nm/us_libya_killings

Greenwald, G. (2011). The Libya War Argument. *Salon*, August 22.

http://www.salon.com/news/opinion/glenn_greenwald/2011/08/22/libya/index.html

Guanaguanare. (2010). The Mighty Sparrow: "I Owe No Apology". *Guanaguanare: The Laughing Gull*, September 15.

http://guanaguanaresingsat.blogspot.com/2010/09/i-owe-no-apology.html

Hamid, S. (2011). Lessons of the Libya Intervention. *The Atlantic*, August 22.
http://www.theatlantic.com/international/archive/2011/08/lessons-of-the-libya-intervention/243922/

Hart, P. (2011a). NYT Points Out 'Racist Overtones' in Libyan Disinformation It Helped Spread. *FAIR–Fairness and Accuracy in Reporting*, August 24.
http://www.fair.org/blog/2011/08/24/nyt-points-out-racist-overtones-in-libyan-disinformation-it-helped-spread/

———. (2011b). Libya and Terrorist Signatures. *FAIR–Fairness and Accuracy in Reporting*, August 30.
http://www.fair.org/blog/2011/08/30/libya-and-terrorist-signatures/

———. (2011b). NYT on WikiLeaks: Move Along, No Atrocity to See Here. *FAIR–Fairness and Accuracy in Reporting*, September 1.
http://www.fair.org/blog/2011/09/01/nyt-on-wikileaks-move-along-no-atrocity-to-see-here/

Harter, P. (2011). Libya Rape Victims "Face Honour Killings". *British Broadcasting Corporation*, June 14.
http://www.bbc.co.uk/news/world-africa-13760895

Hedges, C. (2011). Libya: Here We Go Again. *Truthout*, September 5.
http://www.truthout.com/libya-here-we-go-again/1315225388

Herman, E.S. (2009). Lockerbie in the Propaganda System: Release of Al-Megrahi Evokes Selective History. *Extra! (FAIR–Fairness and Accuracy in Reporting)*, October.
http://www.fair.org/index.php?page=3920

Hill, E. (2011). Migrants Suffer in Tripoli camps. *Al Jazeera English*, September 2.
http://english.aljazeera.net/news/africa/2011/09/201191102134823327.html

Hubbard, B. (2011). Libyan Rebels Round up Black Africans. *Associated Press*, September 1.
http://news.yahoo.com/libyan-rebels-round-black-africans-130723394.html

Human Rights Watch (HRW). (2011). Libya: Opposition Forces Should Protect Civilians and Hospitals—Looting, Arson, and Some Beatings in Captured Western Towns. *Human Rights Watch*, July 13.
http://www.hrw.org/en/news/2011/07/13/libya-opposition-forces-should-protect-civilians-and-hospitals

Hussein, M. (2011). Libya Crisis: What Role Do Tribal Loyalties Play? *British Broadcasting Corporation*, February 21.
http://www.bbc.co.uk/news/mobile/world-middle-east-12528996

IOL. (2011). French Jets Destroy Tanks, Vehicles. *IOL News*, March 19.
http://www.iol.co.za/news/africa/french-jets-destroy-tanks-vehicles-1.1044348

Iqbal, M. (2011a). Libya: The Western-Linked and Backed National Council, and the Hallmarks of War Propaganda. *EmpireStrikesBlack*, March 17.
http://empirestrikesblack.com/2011/03/libya-the-western-linked-and-backed-national-council-and-the-hallmarks-of-war-propaganda/

———. (2011b). Viagra-Induced Rape: ICC Ups the Ante on Propaganda As Desperate NATO Struggles to Crush Libyan Resistance. *EmpireStrikesBlack*, June 8.

http://empirestrikesblack.com/2011/06/rape-viagra-icc-ups-the-ante-on-propaganda-as-desperate-nato-struggles-to-crush-libyan-resistance/

Jacinto, L. (2011). Shifting Loyalties among Libya's Islamists. *France24*, August 10.

http://www.france24.com/en/20110805-libya-uprising-islamists-rebels-ntc-gaddafi-fighters-transition-council-shifting-allies

Kamara, A.M. (2011). Black Africans Executed in Cold-Blood by Rebels in Libya. *The Zimbabwe Mail*, August 26.

http://www.thezimbabwemail.com/zimbabwe/8890-libyan-rebel-ethnic-cleansing-and-lynching-of-black-people.html

Kamat, A., & Shokr, A. (2011). Libya's Reformist Revolutionaries. *Economic & Political Weekly*, 46(12), 13-14.

http://www.box.net/shared/0xd0ll55ro

Keaten, J. (2011). France: Libya Operation May Last Weeks, Not Months. *Associated Press*, March 24.

http://www.cnsnews.com/news/article/france-libya-operation-may-last-weeks-no

Keating, J.E. (2011). Things We Thought We Knew About Libya. *Foreign Policy*, September 1.

http://www.foreignpolicy.com/articles/2011/09/01/things_we_thought_we_knew_about_libya?page=full

Kirkpatrick, D. (2011a). Hopes for a Qaddafi Exit, and Worries of What Comes Next. *The New York Times*, March 21.

http://www.nytimes.com/2011/03/22/world/africa/22tripoli.html?_r=3&ref=daviddkirkpatrick&pagewanted=all

———— . (2011b). Rebels Arm Tripoli Guerrillas and Cut Resources to Capital. *The New York Times*, June 24.

http://www.nytimes.com/2011/06/25/world/africa/25libya.html?_r=2&pagewanted=all

———— . (2011c). NATO Strikes at Libyan State TV. *The New York Times*, July 30.

http://www.nytimes.com/2011/07/31/world/africa/31tripoli.html?_r=2

Kirkpatrick, D., & Nordland, R. (2011). Waves of Disinformation and Confusion Swamp the Truth in Libya. *The New York Times*, August 23.

http://www.nytimes.com/2011/08/24/world/africa/24fog.html?_r=2&pagewanted=all

Kucinich, D. (2011). Time to End Nato's War in Libya. *The Guardian*, August 21.

http://www.guardian.co.uk/commentisfree/cifamerica/2011/aug/21/libya-nato-intervention-gaddafi

Kuperman, A.J. (2011). False pretense for war in Libya? *The Boston Globe*, April 14.

http://articles.boston.com/2011-04-14/bostonglobe/29418371_1_rebel-stronghold-civilians-rebel-positions

Laub, K., & Al-Shalchi, H. (2011). Egypt Protests Leave 297 Killed: Human Rights Watch. *The Huffington Post*, February 8.

http://www.huffingtonpost.com/2011/02/07/egypt-protests-leave-297-_n_819821.html

Laub, K., & Schemm, P. (2011). Libya Rebels Pledge Assault on Gadhafi Stronghold. *Associated Press*, August 30.

http://news.yahoo.com/libya-rebels-pledge-assault-gadhafi-stronghold-133431571.html

Lee, J. (2011). The President on Libya: "The Violence Must Stop; Muammar Gaddafi Has Lost the Legitimacy to Lead and He Must Leave". *The White House Blog*, March 3.

http://www.whitehouse.gov/blog/2011/03/03/president-libya-violence-must-stop-muammar-gaddafi-has-lost-legitimacy-lead-and-he-m

Lekic, S. (2011). EU Set to Relax Libya Sanctions. *Associated Press*, August 31.

http://news.yahoo.com/eu-set-relax-libya-sanctions-132955327.html

MacDougall, C. (2011). How Qaddafi Helped Fuel Fury Toward Africans in Libya. *Christian Science Monitor*, March 6.

http://www.csmonitor.com/World/Africa/2011/0306/How-Qaddafi-helped-fuel-fury-toward-Africans-in-Libya

MacAskill, E. (2011). Gaddafi 'Supplies Troops with Viagra to Encourage Mass Rape', Claims Diplomat. *The Guardian*, April 29.

http://www.guardian.co.uk/world/2011/apr/29/diplomat-gaddafi-troops-viagra-mass-rape

Maclean, W. (2011). RPT–Tripoli's New Normal–Bickering Politicians. *Reuters*, September 14.

http://www.reuters.com/article/2011/09/14/libya-tripoli-idUSL5E7KE08720110914

Malinowski, B. (1926). *Myth in Primitive Psychology*. New York: W.W. Norton & Co.

Martin, P. (2011a). American Media Silent on CIA Ties to Libya Rebel Commander. *World Socialist Web Site*, March 30.

http://wsws.org/articles/2011/mar2011/hift-m30.shtml

——— . (2011b). Mounting Evidence of CIA Ties to Libyan Rebels. *World Socialist Web Site*, April 4.

http://www.wsws.org/articles/2011/apr2011/liby-a04.shtml

Mazetti, M., & Schmitt, E. (2011). C.I.A. Agents in Libya Aid Airstrikes and Meet Rebels. *The New York Times*, March 30.

http://www.nytimes.com/2011/03/31/world/africa/31intel.html?_r=3&smid=tw-nytimesglobal&seid=auto

McCain, J., & Graham, L. (2011). Statement by Senators McCain and Graham on End of the Qaddafi Regime in Libya. *Website of U.S. Senator John McCain of Arizona*, August 21.

http://mccain.senate.gov/public/index.cfm?FuseAction=PressOffice.PressRelease s&ContentRecord_id=ef07da62-0100-107e-d7ac-08531bd793e5

Memoli, M.A. (2011). Tim Pawlenty Brands Obama "Timid" on Libya. *Los Angeles Times*, March 29.

http://articles.latimes.com/2011/mar/29/news/la-pn-republican-reaction-libya-20110330

Moraff, C. (2007). AFRICOM: Round One in a New Cold War? *In These Times*, September 19.

http://www.inthesetimes.com/article/3334/africom_round_one_in_a_new_cold_war/

Mountain, T.C. (2011). 30,000 Bombs over Libya: One Hell of a Humanitarian

Mission. *CounterPunch*, September 2.

http://www.counterpunch.org/2011/09/02/30000-bombs-over-libya/

Mufson, S. (2011). Conflict in Libya: U.S. Oil Companies Sit on Sidelines as Gaddafi Maintains Hold. *The Washington Post*, June 10.

http://www.washingtonpost.com/business/economy/conflict-in-libya-us-oil-companies-sit-on-sidelines-as-gaddafi-maintains-hold/2011/06/03/AGJq2QPH_story.html

Mumisa, M. (2011). Is Al-Jazeera TV Complicit in the Latest Vilification of Libya's Blacks? *The Independent*, February 24.

http://blogs.independent.co.uk/2011/02/24/is-al-jazeerah-tv-complicit-in-the-latest-vilification-of-libyas-blacks/

Murphy, D. (2011). The Gay Girl in Damascus Hoax, 'Mass Rape' in Libya, and Press Credulity Have Our Propaganda Detectors Been Dulled? *Christian Science Monitor*, June 13.

http://www.csmonitor.com/World/Backchannels/2011/0613/The-Gay-Girl-in-Damascus-hoax-mass-rape-in-Libya-and-press-credulity

NATO. (2011a). Statement on Libya. *North Atlantic Treaty Organization*, April 14.

http://www.nato.int/cps/en/natolive/official_texts_72544.htm

———. (2011b). NATO and Libya: Operational Media Update for 5 September. *North Atlantic Treaty Organization*, September 6.

http://www.nato.int/nato_static/assets/pdf/pdf_2011_09/20110906_110906-oup-update.pdf

NBC. (2011). US Intel: No Evidence of Viagra as Weapon in Libya. *MSNBC*, April 29.

http://www.msnbc.msn.com/id/42824884

Nebehay, S. (2011). Rape Used as Weapon of War in Libya and Elsewhere—U.N. *Reuters*, June 10.

http://uk.reuters.com/article/2011/06/10/uk-un-rape-idUKTRE75947120110610

North, D. (2011). Libya, Imperialism and the Prostration of the "Left" Intellectuals: The Case of Professor Juan Cole. *World Socialist Web Site*, April 1.

http://www.wsws.org/articles/2011/apr2011/pers-a01.shtml

Obama, B.H. (2011). Remarks by the President in Address to the Nation on Libya, National Defense University, Washington, D.C., March 28. Washington, DC: Office of the Press Secretary, the White House.

http://www.whitehouse.gov/the-press-office/2011/03/28/remarks-president-address-nation-libya

Obama, B.H., Cameron, D., & Sarkozy, N. (2011). Libya's Pathway to Peace. *International Herald Tribune*, April 14.

http://www.nytimes.com/2011/04/15/opinion/15iht-edlibya15.html

PBS. (2000). War in Europe Facts and Figures. *Frontline*, February 22.

http://www.pbs.org/wgbh/pages/frontline/shows/kosovo/etc/facts.html

Perrin, J-P. (2011). Il y a eu des dizaines de cas de soldats assassins. *Libération*, June 22.

http://www.liberation.fr/monde/01012344751-il-y-a-eu-des-dizaines-de-cas-de-soldats-assassines

Perry, A. (2011). Libyan Leader's Delusions of African Grandeur. *TIME*.
http://www.time.com/time/specials/packages/article/0,28804,2045328_204533
3_2053164,00.html

Pougala, J-P. (2011). Why the West Want the Fall of Muammar Gaddafi. *Tlaxcala*,
April 23.
http://www.tlaxcala-int.org/article.asp?reference=4624

Quist-Arcton, O. (2011). In Libya, African Migrants Say They Face Hostility. *NPR*,
February 25.
http://www.npr.org/2011/02/25/134065767/-African-Migrants-Say-They-Face-
Hostility-From-Libyans

Radio Netherlands Worldwide. (2011). HRW: No Mercenaries in Eastern Libya. *Radio
Netherlands Worldwide*, March 2.
http://www.rnw.nl/africa/article/hrw-no-mercenaries-eastern-libya-0

Raglan, L. (1955). Myth and Ritual. *The Journal of American Folklore*, 68(270), 454-461.

Ramdani, N. (2011). Libya Protests: 'Foreign Mercenaries Using Heavy Weapons
Against At Demonstrators'. *The Telegraph*, February 20.
http://www.telegraph.co.uk/news/worldnews/africaandindianocean/libya/8336
467/Libya-protests-foreign-mercenaries-using-heavy-weapons-against-at-
demonstrators.html

Rasmussen, A.F. (2011a). Statement by the NATO Secretary General on events in
Libya—Press Release (2011) 019. *North Atlantic Treaty Organization*, February 21.
http://www.nato.int/cps/en/natolive/news_70731.htm?mode=pressrelease

———— . (2011b). NATO Secretary General's Statement on the Situation in Libya.
North Atlantic Treaty Organization, February 24.
http://www.nato.int/cps/en/natolive/news_70790.htm?mode=pressrelease

———— . (2011c). @AndersFoghR. *Twitter.com*, February 25.
http://twitter.com/#!/AndersFoghR/status/41034955039588352

———— . (2011d). @AndersFoghR. *Twitter.com*, February 25.
http://twitter.com/#!/AndersFoghR/status/41037040686600192

———— . (2011e). Statement by the NATO Secretary General on the Situation in
Libya—Press Release (2011) 023. *North Atlantic Treaty Organization*, February 25.
http://www.nato.int/cps/en/natolive/news_70893.htm?mode=pressrelease

Reuters. (2011a). UPDATE 1-Wreck of Gaddafi's Force Smoulders Near Benghazi.
Reuters, March 20.
http://af.reuters.com/article/libyaNews/idAFLDE72J0A320110320?sp=true

———— . (2011b). Obama Signed Secret Libya Order Authorizing Support for Rebels.
The Huffington Post, March 30.
http://www.huffingtonpost.com/2011/03/30/obama-secret-order-libya-signed-
rebel-support_n_842734.html

Rice, C. (2006). Secretary Rice Holds a News Conference [Transcript]. *The Washington
Post*, July 21.
http://www.washingtonpost.com/wp-
dyn/content/article/2006/07/21/AR2006072100889.html

———— . (2008). U.S. Secretary of State Condoleezza Rice, Libyan Leader Col
Muammar Abu Minyar al-Qadhafi Discuss Africa Command. *U.S. Africa*

Command, September 6.

http://www.africom.mil/getArticle.asp?art=2026

Richter, P. (2011). U.N. Security Council Authorizes Action against Moammar Kadafi. *Los Angeles Times*, March 18.

http://articles.latimes.com/2011/mar/18/world/la-fg-un-libya-20110318

Sawer, P. (2011). Gaddafi Threatens to Attack Europe Over Airstrikes. *The Telegraph*, July 2.

http://www.telegraph.co.uk/news/worldnews/africaandindianocean/libya/8612344/Gaddafi-threatens-to-attack-Europe-over-airstrikes.html

Schwartz, E. (2011). An American Response to the Arab Spring. *International Affairs Review*, April 19.

http://www.iar-gwu.org/node/305

Schwarz, P. (2011). NATO Allows Libyan Refugees to Drown in the Mediterranean. *World Socialist Web Site*, August 13.

http://www.wsws.org/articles/2011/aug2011/nato-a13.shtml

Scott, P.D. (2011). Who are the Libyan Freedom Fighters and Their Patrons? *GlobalResearch.ca*, March 25.

http://www.globalresearch.ca/index.php?context=va&aid=23947

Sengupta, K. (2011). Rebels Settle Scores in Libyan Capital. *The Independent*, August 27.

http://www.independent.co.uk/news/world/africa/rebels-settle-scores-in-libyan-capital-2344671.html

Shabi, R. (2011). NATO Nations Set to Reap Spoils of Libya War. *Al Jazeera English*, August 25.

http://english.aljazeera.net/indepth/opinion/2011/08/201182511546451332.html

Sheehan, H. (2011). Libya and the Left. *Irish Left Review*, March 9.

http://www.irishleftreview.org/2011/03/09/libya-left/

Shenker, J. (2011). Aircraft Carrier Left Us to Die, Say Migrants. *The Guardian*, May 8.

http://www.guardian.co.uk/world/2011/may/08/nato-ship-libyan-migrants?CMP=twt_gu

Sidner, S. (2011). Libyan Rebels Say Captured Cell Phone Videos Show Rape, Torture. *CNN*, June 17.

http://edition.cnn.com/2011/WORLD/africa/06/14/libya.rape.hfr/index.html

Simpson, V.L. (2011). Italy Demands NATO Probe over Libya Boat Migrants. *Associated Press*, August 5.

http://news.yahoo.com/italy-demands-nato-probe-over-libya-boat-migrants-170944071.html

Skinner, D. (2011). General Ham Discusses First Weeks as AFRICOM Commander at All-Hands Staff Meeting. *U.S. Africa Command*, March 30.

http://www.africom.mil/getArticle.asp?art=6356

Smith, D. (2011). Has Gaddafi Unleashed a Mercenary Force on Libya? *The Guardian*, February 22.

http://www.guardian.co.uk/world/2011/feb/22/gaddafi-mercenary-force-libya

Starr, B. (2011). Foreign Forces in Libya Helping Rebel Forces Advance. *CNN*, August

24.
http://edition.cnn.com/2011/WORLD/africa/08/24/libya.foreign.forces/index
.html?iref=allsearch

Stephen, C. (2011). Muammar Gaddafi War Crimes Files Revealed. *The Observer*, June 18.
http://www.guardian.co.uk/world/2011/jun/18/muammar-gaddafi-war-crimes-files

Stevenson, J. (2011). AFRICOM's Libya Expedition: How War Will Change the Command's Role on the Continent. *Foreign Affairs*, May 9.
http://www.foreignaffairs.com/articles/67844/jonathan-stevenson/africoms-libyan-expedition?page=show

Taki, J. (2011). Libyan Ambassador to UN Urges International Community to Stop Genocide. *Global Arab Network*, February 21.
http://www.english.globalarabnetwork.com/201102219941/Libya-Politics/libyan-ambassador-to-un-urges-international-community-to-stop-genocide.html

Tapper, J., Khan, H., & Raddatz, M. (2011). Obama: U.S. Involvement in Libya Action Would Last "Days, Not Weeks". *ABC News*, March 18.
http://abcnews.go.com/print?id=13164938

Taylor, J. (2011). Dash for Profit in Post-war Libya Carve-Up. *The Independent*, August 24.
http://www.independent.co.uk/news/business/news/dash-for-profit-in-postwar-libya-carveup-2342798.html

Terrill, A. (2011a). The Arab Spring and the Future of U.S. Interests and Cooperative Security in the Arab World. Carlisle, PA: Strategic Studies Institute, U.S. Army War College.
http://www.strategicstudiesinstitute.army.mil/index.cfm/articles/The-Arab-Spring-and-the-Future-of-US-Interests/2011/8/2

———. (2011b). The Arab Upheavals and the Future of the U.S. Military Policies and Presence in the Middle East and the Gulf. Carlisle, PA: Strategic Studies Institute, U.S. Army War College.
http://www.strategicstudiesinstitute.army.mil/index.cfm/articles/Arab-Upheavals-and-the-Future-of-the-US-Military-Policies-and-Presence-in-the-Middle-East-and-the-Gulf/2011/6/27

Thomas, G. (2011). US May Use Covert Action Against Gadhafi. *Voice of America*, March 31.
http://www.voanews.com/english/news/Covert-Action-Might-Target-Gadhafi-117427283.html

Tong, S. (2011). Analysis: Investors Eye Promise, Pitfalls in Post-Gaddafi Libya. *Reuters*, August 22.
http://www.reuters.com/article/2011/08/22/us-libya-investment-idUSTRE77L4NG20110822

United Nations (UN). (1948a). Convention on the Prevention and Punishment of the Crime of Genocide. New York: United Nations General Assembly.
http://www.hrweb.org/legal/genocide.html

———. (1948b). Universal Declaration of Human Rights. Paris: United Nations

General Assembly.

http://www.un.org/en/documents/udhr/

——— . (1949). Convention (III) relative to the Treatment of Prisoners of War. Geneva, 12 August 1949. *ICRC–International Humanitarian Law: Treaties & Documents.*

http://www.icrc.org/ihl.nsf/FULL/375

——— . (1976). International Covenant on Economic, Social and Cultural Rights. New York: United Nations General Assembly.

http://www2.ohchr.org/english/law/cescr.htm

——— . (2005). 2005 World Summit Outcome. New York: United Nations General Assembly.

http://www.un.org/en/preventgenocide/adviser/pdf/World%20Summit%20Outcome%20Document.pdf#page=30

——— . (2006a). Security Council Condemns Attacks against Journalists in Conflict Situations, Unanimously Adopting Resolution 1738 (2006). New York: United Nations Department of Public Information, News and Media Division.

http://www.un.org/News/Press/docs/2006/sc8929.doc.htm

——— . (2006b). Universal Realization of the Right of Peoples to Self-Determination. New York: United Nations General Assembly.

http://www.un.org/ga/search/view_doc.asp?symbol=A/61/333

——— . (2009). Implementing the Responsibility to Protect: Report of the Secretary-General. New York: United Nations General Assembly.

http://www.un.org/ga/search/view_doc.asp?symbol=A/63/677

——— . (2011). Security Council Approves "No-Fly Zone" Over Libya, Authorizing "All Necessary Measures" to Protect Civilians, by Vote of 10 in Favour with 5 Abstentions. New York: United Nations Department of Public Information, News and Media Division.

http://www.un.org/News/Press/docs/2011/sc10200.doc.htm

United Nations Development Programme (UNDP). (2009). *Human Development Report 2009.* New York: United Nations Development Programme.

http://hdr.undp.org/en/media/HDR_2009_EN_Complete.pdf

UN Watch. (2010). Libya Must End Racism Against Black African Migrants and Others (UN Watch Written Statement delivered to the UN Human Rights Council 13th Session). *UN Watch*, February 16.

http://www.unwatch.org/site/apps/nlnet/content2.aspx?c=bdKKISNqEmG&b=1313923&ct=8411733&printmode=1

Upstream. (2011). ENI Chief Mends Libya Fences. *Upstream Online*, August 29.

http://www.upstreamonline.com/live/article275302.ece?WT.mc_id=rechargenews_rss

U.S. Department of Defense. (2011). DOD News Briefing with Secretary Gates and Adm. Mullen from the Pentagon, March 1. Washington, DC: Department of Defense.

http://www.defense.gov/transcripts/transcript.aspx?transcriptid=4777

U.S. Embassy Tripoli. (2007). Growth of Resource Nationalism in Libya. WikiLeaks Libya Cables: Ref ID: 07TRIPOLI967, Date: 11/15/2007, 14:17. Released by WikiLeaks via *The Telegraph*, January 31, 2011.

http://www.telegraph.co.uk/news/wikileaks-files/libya-wikileaks/8294755/GROWTH-OF-RESOURCE-NATIONALISM-IN-LIBYA.html

———— . (2008). Scenesetter for Secretary Rice's Visit to Libya. WikiLeaks Libya Cables: Ref ID: 08TRIPOLI680, Date: 8/29/2008, 18:11. Released by WikiLeaks via *The Telegraph*, January 31, 2011.

http://www.telegraph.co.uk/news/wikileaks-files/libya-wikileaks/8294858/SCENESETTER-FOR-SECRETARY-RICES-VISIT-TO-LIBYA.html

Van Aucken, B. (2011a). An Open Letter to Professor Juan Cole: A Reply to a Slander. *World Socialist Web Site*, August 10.

http://www.wsws.org/articles/2011/aug2011/cole-a10.shtml

———— . (2011b). Professor Cole "Answers" WSWS on Libya: An Admission of Intellectual and Political Bankruptcy. *World Socialist Web Site*, August 16.

http://www.wsws.org/articles/2011/aug2011/cole-a16.shtml

Van Langendonck, G. (2011). In Tripoli, African 'Mercenaries' at Risk. *Christian Science Monitor*, August 29.

http://www.csmonitor.com/World/Middle-East/2011/0829/In-Tripoli-African-mercenaries-at-risk

Varner, B. (2011). Qaddafi May Be Charged With Systematic Rape, Prosecutor Says. *Bloomberg*, June 8.

http://www.bloomberg.com/news/2011-06-08/qaddafi-may-be-charged-with-systematic-rape-prosecutor-says.html

Verkaik, R. (2011). Bombs Hit Gaddafi Home. *Mail Online*, May 1.

http://www.dailymail.co.uk/news/article-1382341/Libya-Nato-strikes-kill-Gaddafis-son-grandchildren.html

Wall Street Journal. (2011). Review and Outlook: The Arab Revolt and U.S. Interests: A U.S. Strategy Has to Begin by Distinguishing Between Friends and Enemies. *Wall Street Journal*, April 4.

http://online.wsj.com/article/SB10001424052748703806304576233103653662410.html

Wallerstein, I. (2011). Libya and the World Left. *Immanuel Wallerstein Commentaries*, March 15.

http://www.iwallerstein.com/libya-world-left/

WSWS. (2011). A CIA Commander for the Libyan Rebels. *World Socialist Web Site*, March 28.

http://www.wsws.org/articles/2011/mar2011/pers-m28.shtml

Zirulnick, A. (2011). ICC: Evidence Shows that Qaddafi Ordered Rape of Hundreds. *Christian Science Monitor*, June 9.

http://www.csmonitor.com/World/Middle-East/2011/0609/ICC-Evidence-shows-that-Qaddafi-ordered-rape-of-hundreds

Žižek, S. (2009). Will the Cat Above the Precipice Fall Down? *The Comment Factory*, June 25.

http://www.thecommentfactory.com/will-the-cat-above-the-precipice-fall-down-slavoj-zizek-on-iran-2259/

Zucchino, D. (2011). Libyan Rebels Appear to Take Leaf from Kadafi's Playbook. *Los*

Angeles Times, March 24.
http://articles.latimes.com/2011/mar/24/world/la-fg-libya-prisoners-20110324

APPENDICES

Appendix A

Farewell Radio and Television Address to the American People. January 17, 1961.
[Delivered from the President's Office at 8:30 p.m.]

By President Dwight D. Eisenhower

My fellow Americans:

Three days from now, after half a century in the service of our country, I shall lay down the responsibilities of office as, in traditional and solemn ceremony, the authority of the Presidency is vested in my successor.

This evening I come to you with a message of leave-taking and farewell, and to share a few final thoughts with you, my countrymen.

Like every other citizen, I wish the new President, and all who will labor with him, Godspeed. I pray that the coming years will be blessed with peace and prosperity for all.

Our people expect their President and the Congress to find essential agreement on issues of great moment, the wise resolution of which will better shape the future of the Nation.

My own relations with the Congress, which began on a remote and tenuous basis when, long ago, a member of the Senate appointed me to

West Point, have since ranged to the intimate during the war and immediate post-war period, and, finally, to the mutually interdependent during these past eight years.

In this final relationship, the Congress and the Administration have, on most vital issues, cooperated well, to serve the national good rather than mere partisanship, and so have assured that the business of the Nation should go forward. So, my official relationship with the Congress ends in a feeling, on my part, of gratitude that we have been able to do so much together.

Dwight D. Eisenhower, White House photo portrait, 29 May 1959. (Source: Wikimedia Commons. In the public domain, under a Creative Commons License.)

II.

We now stand ten years past the midpoint of a century that has witnessed four major wars among great nations. Three of these involved our own country. Despite these holocausts America is today the strongest, the most influential and most productive nation in the world. Understandably proud of this pre-eminence, we yet realize that America's leadership and prestige depend, not merely upon our unmatched material progress, riches and military strength, but on how we use our power in the interests of world peace and human betterment.

III

Throughout America's adventure in free government, our basic purposes have been to keep the peace; to foster progress in human achievement, and to enhance liberty, dignity and integrity among people and among nations. To strive for less would be unworthy of a free and religious people. Any failure traceable to arrogance, or our lack of comprehension or readiness to sacrifice would inflict upon us grievous hurt both at home and abroad.

Progress toward these noble goals is persistently threatened by the conflict now engulfing the world. It commands our whole attention, absorbs our very beings. We face a hostile ideology—global in scope, atheistic in character, ruthless in purpose, and insidious in method. Unhappily the danger is poses promises to be of indefinite duration. To meet it successfully, there is called for, not so much the emotional and transitory sacrifices of crisis, but rather those which enable us to carry forward steadily, surely, and without complaint the burdens of a prolonged and complex struggle—with liberty the stake. Only thus shall we remain, despite every provocation, on our charted course toward permanent peace and human betterment.

Crises there will continue to be. In meeting them, whether foreign or domestic, great or small, there is a recurring temptation to feel that some spectacular and costly action could become the miraculous solution to all current difficulties. A huge increase in newer elements of our defense; development of unrealistic programs to cure every ill in agriculture; a dramatic expansion in basic and applied research—these and many other possibilities, each possibly promising in itself, may be suggested as the only way to the road we wish to travel.

But each proposal must be weighed in the light of a broader consideration: the need to maintain balance in and among national programs—balance between the private and the public economy, balance between cost and hoped for advantage—balance between the clearly necessary and the comfortably desirable; balance between our essential requirements as a nation and the duties imposed by the nation upon the individual; balance between actions of the moment and the national welfare of the future. Good judgment seeks balance and progress; lack of it eventually finds imbalance and frustration.

The record of many decades stands as proof that our people and their government have, in the main, understood these truths and have responded to them well, in the face of stress and threat. But threats, new in kind or degree, constantly arise. I mention two only.

IV

A vital element in keeping the peace is our military establishment. Our arms must be mighty, ready for instant action, so that no potential aggressor may be tempted to risk his own destruction.

Our military organization today bears little relation to that known by any of my predecessors in peacetime, or indeed by the fighting men of World War II or Korea.

Until the latest of our world conflicts, the United States had no armaments industry. American makers of plowshares could, with time and as required, make swords as well. But now we can no longer risk emergency improvisation of national defense; we have been compelled to create a permanent armaments industry of vast proportions. Added to this, three and a half million men and women are directly engaged in the defense establishment. We annually spend on military security more than the net income of all United States corporations.

This conjunction of an immense military establishment and a large arms industry is new in the American experience. The total influence—economic, political, even spiritual—is felt in every city, every State house, every office of the Federal government. We recognize the imperative need for this development. Yet we must not fail to comprehend its grave implications. Our toil, resources and livelihood are all involved; so is the very structure of our society.

In the councils of government, we must guard against the acquisition of unwarranted influence, whether sought or unsought, by the military industrial complex. The potential for the disastrous rise of misplaced power exists and will persist.

We must never let the weight of this combination endanger our liberties or democratic processes. We should take nothing for granted. Only an alert and knowledgeable citizenry can compel the proper meshing of the huge industrial and military machinery of defense with our peaceful methods and goals, so that security and liberty may prosper together.

Akin to, and largely responsible for the sweeping changes in our industrial-military posture, has been the technological revolution during recent decades.

In this revolution, research has become central; it also becomes more formalized, complex, and costly. A steadily increasing share is conducted for, by, or at the direction of, the Federal government.

Today, the solitary inventor, tinkering in his shop, has been overshadowed by task forces of scientists in laboratories and testing fields. In the same fashion, the free university, historically the fountainhead of free ideas and scientific discovery, has experienced a revolution in the conduct of research. Partly because of the huge costs involved, a government contract becomes virtually a substitute for intellectual curiosity. For every old blackboard there are now hundreds of new electronic computers.

The prospect of domination of the nation's scholars by Federal employment, project allocations, and the power of money is ever present and is gravely to be regarded.

Yet, in holding scientific research and discovery in respect, as we should, we must also be alert to the equal and opposite danger that public policy could itself become the captive of a scientific technological elite.

It is the task of statesmanship to mold, to balance, and to integrate these and other forces, new and old, within the principles of our democratic system—ever aiming toward the supreme goals of our free society.

V

Another factor in maintaining balance involves the element of time. As we peer into society's future, we—you and I, and our government—must avoid the impulse to live only for today, plundering, for our own ease and convenience, the precious resources of tomorrow. We cannot mortgage the material assets of our grandchildren without risking the loss also of their political and spiritual heritage. We want democracy to survive for all generations to come, not to become the insolvent phantom of tomorrow.

VI

Down the long lane of the history yet to be written America knows that this world of ours, ever growing smaller, must avoid becoming a community of dreadful fear and hate, and be instead, a proud confederation of mutual trust and respect.

Such a confederation must be one of equals. The weakest must come to the conference table with the same confidence as do we, protected as we are by our moral, economic, and military strength. That table, though scarred by many past frustrations, cannot be abandoned for the certain agony of the battlefield.

Disarmament, with mutual honor and confidence, is a continuing imperative. Together we must learn how to compose differences, not with arms, but with intellect and decent purpose. Because this need is so sharp and apparent I confess that I lay down my official responsibilities in this field with a definite sense of disappointment. As one who has witnessed the horror and the lingering sadness of war—as one who knows that another war could utterly destroy this civilization which has been so slowly and painfully built over thousands of years—I wish I could say tonight that a lasting peace is in sight.

Happily, I can say that war has been avoided. Steady progress toward our ultimate goal has been made. But, so much remains to be done. As a private citizen, I shall never cease to do what little I can to help the world advance along that road.

VII

So—in this my last good night to you as your President—I thank you for the many opportunities you have given me for public service in war and peace. I trust that in that service you find some things worthy; as for the rest of it, I know you will find ways to improve performance in the future.

You and I—my fellow citizens—need to be strong in our faith that all nations, under God, will reach the goal of peace with justice. May we be ever unswerving in devotion to principle, confident but humble with power, diligent in pursuit of the Nation's great goals.

To all the peoples of the world, I once more give expression to America's prayerful and continuing aspiration:

We pray that peoples of all faiths, all races, all nations, may have their great human needs satisfied; that those now denied opportunity shall come to enjoy it to the full; that all who yearn for freedom may experience its spiritual blessings; that those who have freedom will understand, also, its heavy responsibilities; that all who are insensitive to the needs of others will learn charity; that the scourges of poverty, disease and ignorance will be made to disappear from the earth, and that, in the goodness of time, all peoples will come to live together in a peace guaranteed by the binding force of mutual respect and love.

Appendix B

War is a Racket

By Major General Smedley Butler

Round Table Press, Inc., 1935
Collection: opensource
http://www.archive.org/details/WarIsARacket

CHAPTER ONE
War Is a Racket

W AR is a racket. It always has been.
It is possibly the oldest, easily the most profitable, surely the most vicious. It is the only one international in scope. It is the only one in which the profits are reckoned in dollars and the losses in lives.

A racket is best described, I believe, as something that is not what it seems to the majority of the people. Only a small "inside" group knows what it is about. It is conducted for the benefit of the very few, at the expense of the very many. Out of war a few people make huge fortunes.

In the World War [I] a mere handful garnered the profits of the conflict. At least 21,000 new millionaires and billionaires were made in the United States during the World War. That many admitted their huge blood gains in their income tax returns. How many other war millionaires falsified their tax returns no one knows.

How many of these war millionaires shouldered a rifle? How many of them dug a trench? How many of them knew what it meant to go hungry in a rat-infested dug-out? How many of them spent sleepless, frightened nights, ducking shells and shrapnel and machine gun bullets? How many of them parried a bayonet thrust of an enemy? How many of them were wounded or killed in battle?

Major General Smedley Butler. (Source: Wikimedia Commons. In the public domain, under a Creative Commons License.)

Out of war nations acquire additional territory, if they are victorious. They just take it. This newly acquired territory promptly is exploited by the few—the selfsame few who wrung dollars out of blood in the war. The general public shoulders the bill.

And what is this bill?

This bill renders a horrible accounting. Newly placed gravestones. Mangled bodies. Shattered minds. Broken hearts and homes. Economic instability. Depression and all its attendant miseries. Back-breaking taxation for generations and generations.

For a great many years, as a soldier, I had a suspicion that war was a racket; not until I retired to civil life did I fully realize it. Now that I see the international war clouds gathering, as they are today, I must face it and speak out.

Again they are choosing sides. France and Russia met and agreed to stand side by side. Italy and Austria hurried to make a similar agreement. Poland and Germany cast sheep's eyes at each other, forgetting for the nonce [one unique occasion], their dispute over the Polish Corridor.

The assassination of King Alexander of Jugoslavia [Yugoslavia] complicated matters. Jugoslavia and Hungary, long bitter enemies, were almost at each other's throats. Italy was ready to jump in. But France was waiting. So was Czechoslovakia. All of them are looking ahead to war. Not the people—not those who fight and pay and die—only those who foment wars and remain safely at home to profit.

There are 40,000,000 men under arms in the world today, and our statesmen and diplomats have the temerity to say that war is not in the making.

Hell's bells! Are these 40,000,000 men being trained to be dancers?

Not in Italy, to be sure. Premier Mussolini knows what they are being trained for. He, at least, is frank enough to speak out. Only the other day, Il Duce in "International Conciliation," the publication of the Carnegie Endowment for International Peace, said:

> "And above all. Fascism, the more it considers and observes the future and the development of humanity quite apart from political considerations of the moment, believes neither in the possibility nor the utility of perpetual peace....War alone brings up to its highest tension all human energy and puts the stamp of nobility upon the people who have the courage to meet it".

Undoubtedly Mussolini means exactly what he says. His well-trained army, his great fleet of planes, and even his navy are ready for war—anxious for it, apparently. His recent stand at the side of Hungary in the latter's dispute with Jugoslavia showed that. And the hurried mobilization of his troops on the Austrian border after the assassination of Dollfuss showed it too. There are others in Europe too whose sabre rattling presages war, sooner or later.

Herr Hitler, with his rearming Germany and his constant demands for more and more arms, is an equal if not greater menace to peace. France only recently increased the term of military service for its youth from a year to eighteen months.

Yes, all over, nations are camping in their arms. The mad dogs of Europe are on the loose. In the Orient the maneuvering is more adroit. Back in 1904, when Russia and Japan fought, we kicked out our old friends the Russians and backed Japan. Then our very generous international bankers were financing Japan. Now the trend is to poison us against the Japanese. What does the "open door" policy to China mean to us? Our trade with China is about $90,000,000 a year. Or the Philippine Islands? We have spent about $600,000,000 in the Philippines in thirty-five years and we (our bankers and industrialists and speculators) have private investments there of less than $200,000,000.

Then, to save that China trade of about $90,000,000, or to protect these private investments of less than $200,000,000 in the Philippines, we would be all stirred up to hate Japan and go to war—a war that might well cost us tens of billions of dollars, hundreds of thousands of lives of Americans, and many more hundreds of thousands of physically maimed and mentally unbalanced men.

Of course, for this loss, there would be a compensating profit—fortunes would be made. Millions and billions of dollars would be piled up. By a few. Munitions makers. Bankers. Ship builders. Manufacturers. Meat packers. Speculators. They would fare well.

Yes, they are getting ready for another war. Why shouldn't they? It pays high dividends.

But what does it profit the men who are killed? What does it profit their mothers and sisters, their wives and their sweethearts? What does it profit their children?

What does it profit anyone except the very few to whom war means huge profits?

Yes, and what does it profit the nation?

Take our own case. Until 1898 we didn't own a bit of territory outside the mainland of North America. At that time our national debt was a little more than $1,000,000,000. Then we became "internationally minded". We forgot, or shunted aside, the advice of the Father of our country. We forgot George Washington's warning about "entangling alliances". We went to war. We acquired outside territory. At the end of the World War period, as a direct result of our fiddling in international affairs, our national debt had jumped to over $25,000,000,000. Our total favorable trade balance during the twenty-five-year period was about $24,000,000,000. Therefore, on a purely bookkeeping basis, we ran a little behind year for year, and that foreign trade might well have been ours without the wars.

It would have been far cheaper (not to say safer) for the average American who pays the bills to stay out of foreign entanglements. For a very few this racket, like bootlegging and other underworld rackets, brings fancy profits, but the cost of operations is always transferred to the people—who do not profit.

CHAPTER TWO
Who Makes The Profits?

The World War, rather our brief participation in it, has cost the United States some $52,000,000,000. Figure it out. That means $400 to every American man, woman, and child. And we haven't paid the debt yet. We are paying it, our children will pay it, and our children's children probably still will be paying the cost of that war.

The normal profits of a business concern in the United States are six, eight, ten, and sometimes twelve percent. But war-time profits—ah! that is another matter—twenty, sixty, one hundred, three hundred, and even eighteen hundred per cent—the sky is the limit. All that traffic will bear. Uncle Sam has the money. Let's get it.

Of course, it isn't put that crudely in war time. It is dressed into speeches about patriotism, love of country, and "we must all put our shoulders to the wheel," but the profits jump and leap and skyrocket— and are safely pocketed. Let's just take a few examples:

Take our friends the du Fonts, the powder people—didn't one of them testify before a Senate committee recently that their powder won the war? Or saved the world for democracy? Or something? How did they do in the war? They were a patriotic corporation. Well, the average earnings of the du Fonts for the period 1910 to 1914 were $6,000,000 a year. It wasn't much, but the du Fonts managed to get along on it. Now let's look at their average yearly profit during the war years, 1914 to 1918. Fifty-eight million dollars a year profit we find! Nearly ten times that of normal times, and the profits of normal times were pretty good. An increase in profits of more than 950 per cent.

Take one of our little steel companies that patriotically shunted aside the making of rails and girders and bridges to manufacture war materials. Well, their 1910-1914 yearly earnings averaged $6,000,000.

Then came the war. And, like loyal citizens, Bethlehem Steel promptly turned to munitions making. Did their profits jump—or did they let Uncle Sam in for a bargain? Well, their 1914-1918 average was $49,000,000 a year!

Or, let's take United States Steel. The normal earnings during the five-year period prior to the war were $105,000,000 a year. Not bad. Then along came the war and up went the profits. The average yearly profit for the period 1914-1918 was $240,000,000. Not bad.

There you have some of the steel and powder earnings. Let's look at something else. A little copper, perhaps. That always does well in war times.

Anaconda, for instance. Average yearly earnings during the pre-war years 1910-1914 of $10,000,000. During the war years 1914-1918 profits leaped to $34,000,000 per year.

Or Utah Copper. Average of $5,000,000 per year during the 1910-1914 period. Jumped to an average of $21,000,000 yearly profits for the war period.

Let's group these five, with three smaller companies. The total yearly average profits of the pre-war period 1910-1914 were $137,480,000. Then along came the war. The average yearly profits for this group sky-rocketed to $408,300,000.

A little increase in profits of approximately 200 per cent.

Does war pay? It paid them. But they aren't the only ones. There are still others. Let's take leather.

For the three-year period before the war the total profits of Central Leather Company were $3,500,000. That was approximately $1,167,000 a year. Well, in 1916 Central Leather returned a profit of $15,000,000, a small increase of 1,100 per cent. That's all. The General Chemical Company averaged a profit for the three years before the war of a little over $800,000 a year. Came the war, and the profits jumped to $12,000,000. a leap of 1,400 per cent.

International Nickel Company—and you can't have a war without nickel—showed an increase in profits from a mere average of $4,000,000 a year to $73,000,000 yearly. Not bad? An increase of more than 1,700 per cent.

American Sugar Refining Company averaged $2,000,000 a year for the three years before the war. In 1916 a profit of $6,000,000 was recorded.

Listen to Senate Document No. 259. The Sixty-Fifth Congress, reporting on corporate earnings and government revenues. Considering the profits of 122 meat packers, 153 cotton manufacturers, 299 garment makers, 49 steel plants, and 340 coal producers during the war.

Profits under 25 per cent were exceptional. For instance the coal companies made between 100 per cent and 7,856 per cent on their capital stock during the war. The Chicago packers doubled and tripled their earnings.

And let us not forget the bankers who financed the great war. If anyone had the cream of the profits it was the bankers. Being partnerships rather than incorporated organizations, they do not have to report to stockholders. And their profits were as secret as they were immense. How the bankers made their millions and their billions I do not know, because those little secrets never become public—even before a Senate investigatory body.

But here's how some of the other patriotic industrialists and speculators chiseled their way into war profits.

Take the shoe people. They like war. It brings business with abnormal profits. They made huge profits on sales abroad to our allies. Perhaps, like the munitions manufacturers and armament makers, they also sold to the enemy. For a dollar is a dollar whether it comes from Germany or from France. But they did well by Uncle Sam too. For instance, they sold Uncle Sam 35,000,000 pairs of hobnailed service shoes. There were 4,000,000 soldiers. Eight pairs, and more, to a soldier. My regiment during the war had only one pair to a soldier. Some of these shoes probably are still in existence. They were good shoes. But when the war was over Uncle Sam has a matter of 25,000,000 pairs left over. Bought—and paid for. Profits recorded and pocketed.

There was still lots of leather left. So the leather people sold your Uncle Sam hundreds of thousands of McClellan saddles for the cavalry. But there wasn't any American cavalry overseas! Somebody had to get rid of this leather, however. Somebody had to make a profit in it—so we had a lot of McClellan saddles. And we probably have those yet.

Also somebody had a lot of mosquito netting. They sold your Uncle Sam 20,000,000 mosquito nets for the use of the soldiers overseas. I suppose the boys were expected to put it over them as they tried to sleep in muddy trenches—one hand scratching cooties on their backs and the other making passes at scurrying rats. Well, not one of these mosquito nets ever got to France!

Anyhow, these thoughtful manufacturers wanted to make sure that no soldier would be without his mosquito net, so 40,000,000 additional yards of mosquito netting were sold to Uncle Sam.

There were pretty good profits in mosquito netting in those days, even if there were no mosquitoes in France. I suppose, if the war had lasted just a little longer, the enterprising mosquito netting manufacturers would have sold your Uncle Sam a couple of consignments of mosquitoes to plant in France so that more mosquito netting would be in order.

Airplane and engine manufacturers felt they, too, should get their just profits out of this war. Why not? Everybody else was getting theirs. So $1,000,000,000—count them if you live long enough—was spent by Uncle Sam in building airplane engines that never left the ground! Not one plane, or motor, out of the billion dollars worth ordered, ever got into a battle in France. Just the same the manufacturers made their little profit of 30, 100, or perhaps 300 per cent.

Undershirts for soldiers cost 140 [cents] to make and uncle Sam paid 300 to 400 each for them—a nice little profit for the undershirt manufacturer. And the stocking manufacturer and the uniform manufacturers and the cap manufacturers and the steel helmet manufacturers—all got theirs.

Why, when the war was over some 4,000,000 sets of equipment—knapsacks and the things that go to fill them—crammed warehouses on this side. Now they are being scrapped because the regulations have changed the contents. But the manufacturers collected their wartime profits on them—and they will do it all over again the next time.

There were lots of brilliant ideas for profit making during the war.

One very versatile patriot sold Uncle Sam twelve dozen 48-inch wrenches. Oh, they were very nice wrenches. The only trouble was that there was only one nut ever made that was large enough for these wrenches. That is the one that holds the turbines at Niagara Falls. Well, after Uncle Sam had bought them and the manufacturer had pocketed the profit, the wrenches were put on freight cars and shunted all around the United States in an effort to find a use for them. When the Armistice was signed it was indeed a sad blow to the wrench manufacturer. He was just about to make some nuts to fit the wrenches. Then he planned to sell these, too, to your Uncle Sam.

Still another had the brilliant idea that colonels shouldn't ride in automobiles, nor should they even ride on horseback. One has probably seen a picture of Andy Jackson riding in a buckboard. Well, some 6,000 buckboards were sold to Uncle Sam for the use of colonels! Not one of them was used. But the buckboard manufacturer got his war profit.

The shipbuilders felt they should come in on some of it, too. They built a lot of ships that made a lot of profit. More than $3,000,000,000 worth. Some of the ships were all right. But $635,000,000 worth of them were made of wood and wouldn't float! The seams opened up—and they sank. We paid for them, though. And somebody pocketed the profits.

It has been estimated by statisticians and economists and research-ers that the war cost your Uncle Sam $52,000,000,000. Of this sum, $39,000,000,000 was expended in the actual war itself. This expendi-ture yielded $16,000,000,000 in profits. That is how the 21,000 billion-aires and millionaires got that way. This $16,000,000,000 profits is not to be sneezed at. It is quite a tidy sum. And it went to a very few.

The Senate (Nye) committee probe of the munitions industry and its wartime profits, despite its sensational disclosures, hardly has scratched the surface.

Even so, it has had some effect. The State Department has been studying "for some time" methods of keeping out of war. The War De-partment suddenly decides it has a wonderful plan to spring. The Ad-ministration names a committee—with the War and Navy Departments ably represented under the chairmanship of a Wall Street speculator—to limit profits in war time. To what extent isn't suggested. Hmmm. Possi-bly the profits of 300 and 600 and 1,600 per cent of those who turned blood into gold in the World War would be limited to some smaller figure.

Apparently, however, the plan does not call for any limitation of losses—that is, the losses of those who fight the war. As far as I have been able to ascertain there is nothing in the scheme to limit a soldier to the loss of but one eye, or one arm, or to limit his wounds to one or two or three. Or to limit the loss of life.

There is nothing in this scheme, apparently, that says not more than 12 per cent of a regiment shall be wounded in battle, or that not more than 7 per cent in a division shall be killed.

Of course, the committee cannot be bothered with such trifling matters.

CHAPTER THREE
Who Pays The Bills?

Who provides the profits—these nice little profits of 20, 100, 300, 1,500 and 1,800 per cent? We all pay them—in taxation. We paid the bankers their profits when we bought Liberty Bonds at $100.00 and sold them back at $84 or $86 to the bankers. These bankers collected $100 plus. It was a simple manipulation. The bankers control the security marts. It was easy for them to depress the price of these bonds. Then all of us—the people—got frightened and sold the bonds at $84 or $86. The bank-ers bought them. Then these same bankers stimulated a boom and government bonds went to par—and above. Then the bankers collected their profits.

But the soldier pays the biggest part of the bill.

If you don't believe this, visit the American cemeteries on the battle-fields abroad. Or visit any of the veteran's hospitals in the United States. On a tour of the country, in the midst of which I am at the time of this writing, I have visited eighteen government hospitals for veterans. In them are a total of about 50,000 destroyed men—men who were the pick of the nation eighteen years ago. The very able chief surgeon at the government hospital; at Milwaukee, where there are 3,800 of the living dead, told me that mortality among veterans is three times as great as among those who stayed at home.

Boys with a normal viewpoint were taken out of the fields and offices and factories and classrooms and put into the ranks. There they were remolded; they were made over; they were made to "about face"; to regard murder as the order of the day. They were put shoulder to shoulder and, through mass psychology, they were entirely changed. We used them for a couple of years and trained them to think nothing at all of killing or of being killed.

Then, suddenly, we discharged them and told them to make another "about face" ! This time they had to do their own readjustment, sans [without] mass psychology, sans officers' aid and advice and sans nation-wide propaganda. We didn't need them any more. So we scattered them about without any "three-minute" or "Liberty Loan" speeches or parades. Many, too many, of these fine young boys are eventually destroyed, mentally, because they could not make that final "about face" alone.

In the government hospital in Marion, Indiana, 1,800 of these boys are in pens! Five hundred of them in a barracks with steel bars and wires all around outside the buildings and on the porches. These already have been mentally destroyed. These boys don't even look like human beings. Oh, the looks on their faces! Physically, they are in good shape; mentally, they are gone.

There are thousands and thousands of these cases, and more and more are coming in all the time. The tremendous excitement of the war, the sudden cutting off of that excitement—the young boys couldn't stand it.

That's a part of the bill. So much for the dead—they have paid their part of the war profits. So much for the mentally and physically wounded—they are paying now their share of the war profits. But the others paid, too—they paid with heartbreaks when they tore themselves away from their firesides and their families to don the uniform of Uncle

Sam—on which a profit had been made. They paid another part in the training camps where they were regimented and drilled while others took their jobs and their places in the lives of their communities. The paid for it in the trenches where they shot and were shot; where they were hungry for days at a time; where they slept in the mud and the cold and in the rain—with the moans and shrieks of the dying for a horrible lullaby.

But don't forget—the soldier paid part of the dollars and cents bill too.

Up to and including the Spanish-American War, we had a prize system, and soldiers and sailors fought for money. During the Civil War they were paid bonuses, in many instances, before they went into service. The government, or states, paid as high as $1,200 for an enlistment. In the Spanish-American War they gave prize money. When we captured any vessels, the soldiers all got their share—at least, they were supposed to. Then it was found that we could reduce the cost of wars by taking all the prize money and keeping it, but conscripting [drafting] the soldier anyway. Then soldiers couldn't bargain for their labor. Everyone else could bargain, but the soldier couldn't.

Napoleon once said,

"All men are enamored of decorations . . . they positively hunger for them".

So by developing the Napoleonic system—the medal business—the government learned it could get soldiers for less money, because the boys liked to be decorated. Until the Civil War there were no medals. Then the Congressional Medal of Honor was handed out. It made enlistments easier. After the Civil War no new medals were issued until the Spanish- American War.

In the World War, we used propaganda to make the boys accept conscription. They were made to feel ashamed if they didn't join the army.

So vicious was this war propaganda that even God was brought into it. With few exceptions our clergymen joined in the clamor to kill, kill, kill. To kill the Germans. God is on our side...it is His will that the Germans be killed.

And in Germany, the good pastors called upon the Germans to kill the allies...to please the same God. That was a part of the general propaganda, built up to make people war conscious and murder conscious.

Beautiful ideals were painted for our boys who were sent out to die. This was the "war to end all wars". This was the "war to make the world safe for democracy". No one mentioned to them, as they marched away, that their going and their dying would mean huge war profits. No one told these American soldiers that they might be shot down by bullets made by their own brothers here. No one told them that the ships on which they were going to cross might be torpedoed by submarines built with United States patents. They were just told it was to be a "glorious adventure".

Thus, having stuffed patriotism down their throats, it was decided to make them help pay for the war, too. So, we gave them the large salary of $30 a month.

All they had to do for this munificent sum was to leave their dear ones behind, give up their jobs, lie in swampy trenches, eat canned willy (when they could get it) and kill and kill and kill...and be killed.

But wait!

Half of that wage (just a little more than a riveter in a shipyard or a laborer in a munitions factory safe at home made in a day) was promptly taken from him to support his dependents, so that they would not become a charge upon his community. Then we made him pay what amounted to accident insurance—something the employer pays for in an enlightened state—and that cost him $6 a month. He had less than $9 a month left.

Then, the most crowning insolence of all—he was virtually black-jacked into paying for his own ammunition, clothing, and food by being made to buy Liberty Bonds. Most soldiers got no money at all on pay days.

We made them buy Liberty Bonds at $100 and then we bought them back—when they came back from the war and couldn't find work—at $84 and $86. And the soldiers bought about $2,000,000,000 worth of these bonds!

Yes, the soldier pays the greater part of the bill. His family pays too. They pay it in the same heart-break that he does. As he suffers, they suffer. At nights, as he lay in the trenches and watched shrapnel burst about him, they lay home in their beds and tossed sleeplessly—his father, his mother, his wife, his sisters, his brothers, his sons, and his daughters.

When he returned home minus an eye, or minus a leg or with his mind broken, they suffered too—as much as and even sometimes more than he. Yes, and they, too, contributed their dollars to the profits of the munitions makers and bankers and shipbuilders and the manufacturers and the speculators made. They, too, bought Liberty Bonds and contributed to the profit of the bankers after the Armistice in the hocus-pocus of manipulated Liberty Bond prices.

And even now the families of the wounded men and of the mentally broken and those who never were able to readjust themselves are still suffering and still paying.

CHAPTER FOUR
How to Smash This Racket!

WELL, it's a racket, all right.

A few profit—and the many pay. But there is a way to stop it. You can't end it by disarmament conferences. You can't eliminate it by peace parleys at Geneva. Well-meaning but impractical groups can't wipe it out by resolutions. It can be smashed effectively only by taking the profit out of war.

The only way to smash this racket is to conscript capital and industry and labor before the nations manhood can be conscripted. One month before the Government can conscript the young men of the nation—it must conscript capital and industry and labor. Let the officers and the directors and the high-powered executives of our armament factories and our munitions makers and our shipbuilders and our airplane builders and the manufacturers of all the other things that provide profit in war time as well as the bankers and the speculators, be conscripted—to get $30 a month, the same wage as the lads in the trenches get.

Let the workers in these plants get the same wages—all the workers, all presidents, all executives, all directors, all managers, all bankers—yes, and all generals and all admirals and all officers and all politicians and all government office holders—everyone in the nation be restricted to a total monthly income not to exceed that paid to the soldier in the trenches!

Let all these kings and tycoons and masters of business and all those workers in industry and all our senators and governors and majors pay half of their monthly $30 wage to their families and pay war risk insurance and buy Liberty Bonds.

Why shouldn't they?

They aren't running any risk of being killed or of having their bodies mangled or their minds shattered. They aren't sleeping in muddy trenches. They aren't hungry. The soldiers are!

Give capital and industry and labor thirty days to think it over and you will find, by that time, there will be no war. That will smash the war racket—that and nothing else.

Maybe I am a little too optimistic. Capital still has some say. So capital won't permit the taking of the profit out of war until the people—those who do the suffering and still pay the price—make up their minds that those they elect to office shall do their bidding, and not that of the profiteers.

Another step necessary in this fight to smash the war racket is the limited plebiscite to determine whether a war should be declared. A plebiscite not of all the voters but merely of those who would be called upon to do the fighting and dying. There wouldn't be very much sense in having a 76-year-old president of a munitions factory or the flat-footed head of an international banking firm or the cross-eyed manager of a uniform manufacturing plant—all of whom see visions of tremendous profits in the event of war—voting on whether the nation should go to war or not. They never would be called upon to shoulder arms—to sleep in a trench and to be shot. Only those who would be called upon to risk their lives for their country should have the privilege of voting to determine whether the nation should go to war.

There is ample precedent for restricting the voting to those affected. Many of our states have restrictions on those permitted to vote. In most, it is necessary to be able to read and write before you may vote. In some, you must own property. It would be a simple matter each year for the men coming of military age to register in their communities as they did in the draft during the World War and be examined physically. Those who could pass and who would therefore be called upon to bear arms in the event of war would be ehgible to vote in a limited plebiscite. They should be the ones to have the power to decide—and not a Congress few of whose members are within the age limit and fewer still of whom are in physical condition to bear arms. Only those who must suffer should have the right to vote.

A third step in this business of smashing the war racket is to make certain that our military forces are truly forces for defense only.

At each session of Congress the question of further naval appropriations comes up. The swivel-chair admirals of Washington (and there are always a lot of them) are very adroit lobbyists. And they are smart. They don't shout that "We need a lot of battleships to war on this nation or that nation". Oh no. First of all, they let it be known that America is menaced by a great naval power. Almost any day, these admirals will tell you, the great fleet of this supposed enemy will strike suddenly and annihilate 125,000,000 people. Just like that. Then they begin to cry for a larger navy. For what? To fight the enemy? Oh my, no. Oh, no. For defense purposes only.

Then, incidentally, they announce maneuvers in the Pacific. For defense. Uh, huh.

The Pacific is a great big ocean. We have a tremendous coastline on the Pacific. Will the maneuvers be off the coast, two or three hundred miles? Oh, no. The maneuvers will be two thousand, yes, perhaps even thirty-five hundred miles, off the coast.

The Japanese, a proud people, of course will be pleased beyond expression to see the United States fleet so close to Nippon's shores. Even as pleased as would be the residents of California were they to dimly discern through the morning mist, the Japanese fleet playing at war games off Los Angeles.

The ships of our navy, it can be seen, should be specifically limited, by law, to within 200 miles of our coastline. Had that been the law in 1898 the Maine would never have gone to Havana Harbor. She never would have been blown up. There would have been no war with Spain with its attendant loss of life. Two hundred miles is ample, in the opinion of experts, for defense purposes. Our nation cannot start an offensive war if its ships can't go further than 200 miles from the coastline. Planes might be permitted to go as far as 500 miles from the coast for purposes of reconnaissance. And the army should never leave the territorial limits of our nation.

To summarize: Three steps must be taken to smash the war racket.

1. We must take the profit out of war.

2. We must permit the youth of the land who would bear arms to decide whether or not there should be war.

3. We must limit our military forces to home defense purposes.

CHAPTER FIVE
To Hell with War!

I am not a fool as to believe that war is a thing of the past. I know the people do not want war, but there is no use in saying we cannot be pushed into another war.

Looking back, Woodrow Wilson was re-elected president in 1916 on a platform that he had "kept us out of war" and on the implied promise that he would "keep us out of war". Yet, five months later he asked Congress to declare war on Germany.

In that five-month interval the people had not been asked whether they had changed their minds. The 4,000,000 young men who put on uniforms and marched or sailed away were not asked whether they wanted to go forth to suffer and die.

Then what caused our government to change its mind so suddenly? Money.

An allied commission, it may be recalled, came over shortly before the war declaration and called on the President. The President summoned a group of advisers. The head of the commission spoke. Stripped of its diplomatic language, this is what he told the President and his group:

> "There is no use kidding ourselves any longer. The cause of the allies is lost. We now owe you (American bankers, American munitions makers, American manufacturers, American speculators, American exporters) five or six billion dollars.

> "If we lose (and without the help of the United States we must lose) we, England, France and Italy, cannot pay back this money...and Germany won't.

> So ..."

Had secrecy been outlawed as far as war negotiations were concerned, and had the press been invited to be present at that conference, or had radio been available to broadcast the proceedings, America never would have entered the World War. But this conference, like all war discussions, was shrouded in utmost secrecy. When our boys were sent off to war they were told it was a "war to make the world safe for democracy" and a "war to end all wars".

Well, eighteen years after, the world has less of democracy than it had then. Besides, what business is it of ours whether Russia or Germany or England or France or Italy or Austria live under democracies or monarchies? Whether they are Fascists or Communists? Our problem is to preserve our own democracy.

And very little, if anything, has been accomplished to assure us that the World War was really the war to end all wars.

Yes, we have had disarmament conferences and limitations of arms conferences. They don't mean a thing. One has just failed; the results of another have been nullified. We send our professional soldiers and our sailors and our politicians and our diplomats to these conferences. And what happens?

The professional soldiers and sailors don't want to disarm. No admiral wants to be without a ship. No general wants to be without a command. Both mean men without jobs. They are not for disarmament. They cannot be for limitations of arms. And at all these conferences, lurking in the background but all-powerful, just the same, are the sinister agents of those who profit by war. They see to it that these conferences do not disarm or seriously limit armaments.

The chief aim of any power at any of these conferences has not been to achieve disarmament to prevent war but rather to get more armament for itself and less for any potential foe.

There is only one way to disarm with any semblance of practicability. That is for all nations to get together and scrap every ship, every gun, every rifle, every tank, every war plane. Even this, if it were possible, would not be enough.

The next war, according to experts, will be fought not with battleships, not by artillery, not with rifles and not with machine guns. It will be fought with deadly chemicals and gases.

Secretly each nation is studying and perfecting newer and ghastlier means of annihilating its foes wholesale. Yes, ships will continue to be built, for the shipbuilders must make their profits. And guns still will be manufactured and powder and rifles will be made, for the munitions makers must make their huge profits. And the soldiers, of course, must wear uniforms, for the manufacturer must make their war profits too.

But victory or defeat will be determined by the skill and ingenuity of our scientists.

If we put them to work making poison gas and more and more fiendish mechanical and explosive instruments of destruction, they will have no time for the constructive job of building greater prosperity for all peoples. By putting them to this useful job, we can all make more money out of peace than we can out of war—even the munitions makers.

So. ..I say,

TO HELL WITH WAR!

Appendix C

The War Prayer

By Mark Twain

It was a time of great exulting and excitement. The country was up in arms, the war was on, in every breast burned the holy fire of patriotism; the drums were beating, the bands playing, the toy pistols popping, the bunched firecrackers hissing and sputtering; on every hand and far down the receding and fading spread of roofs and balconies a fluttering wilderness of flags flashed in the sun; daily the young volunteers marched down the wide avenue gay and fine in their new uniforms, the proud fathers and mothers and sisters and sweethearts cheering them with voices choked with happy emotion as they swung by; nightly the packed mass meetings listened, panting, to patriot oratory which stirred the deepest depths of their hearts, and which they interrupted at briefest intervals with cyclones of applause, the tears running down their cheeks the while; in the churches the pastors preached devotion to flag and country, and invoked the God of Battles, beseeching His aid in our good cause in outpourings of fervid eloquence which moved every listener. It was indeed a glad and gracious time, and the half dozen rash spirits that ventured to disapprove of the war and cast doubt upon its righteousness straight way got such a stern and angry warning that for their personal safety's sake they quickly shrank out of sight and offended no more in that way.

Sunday morning came—next day the battalions would leave for the front; the church was filled; the volunteers were there, their young faces alight with martial dreams—visions of the stern advance, the gathering momentum, the rushing charge, the flashing sabers, the flight of the foe, the tumult, the enveloping smoke, the fierce pursuit, the surrender!—then home from the war, bronzed heroes, welcomed, adored, submerged in golden seas of glory! With the volunteers sat their dear ones, proud, happy, and envied by the neighbors and friends who had no sons and brothers to send forth to the field of honor, there to win for the flag, or failing, die the noblest of noble deaths. The service proceeded; a war chapter from the Old Testament was read; the first prayer was said; it was followed by an organ burst that shook the building, and with one impulse the house rose, with glowing eyes and beating hearts, and poured out that tremendous invocation:

A portrait of Mark Twain taken by A. F. Bradley in New York, 1907. (Source: Wikimedia Commons. In the public domain.)

"God the all-terrible! Thou who ordainest, Thunder thy clarion and lightning thy sword!"

Then came the "long" prayer. None could remember the like of it for passionate pleading and moving and beautiful language. The burden of its supplication was, that an ever-merciful and benignant Father of us all would watch over our noble young soldiers, and aid, comfort, and encourage them in their patriotic work; bless them, shield them in the day of battle and the hour of peril, bear them in His mighty hand, make them strong and confident, invincible in the bloody onset; help them to crush the foe, grant to them and to their flag and country imperishable honor and glory—An aged stranger entered and moved with slow and noiseless step up the main aisle, his eyes fixed upon the minister, his long body clothed in a robe that reached to his feet, his head bare, his white hair descending in a frothy cataract to his shoulders, his seamy face unnaturally pale, pale even to ghastliness. With all eyes following and wondering, he made his silent way; without pausing, he ascended to the preacher's side and stood there, waiting. With shut lids the preacher, unconscious of his presence, continued his moving prayer, and at last finished it with the words, uttered in fervent appeal, "Bless our arms, grant us victory, O Lord our God, Father and Protector of our land and flag!"

The stranger touched his arm, motioned him to step aside—which the startled minister did—and took his place. During some moments he surveyed the spellbound audience with solemn eyes, in which burned an uncanny light; then in a deep voice he said:

"I come from the Throne—bearing a message from Almighty God!" The words smote the house with a shock; if the stranger perceived it he gave no attention. "He has heard the prayer of His servant your shepherd, and will grant it if such be your desire after I, His messenger, shall have explained to you its import—that is to say, its full import. For it is like unto many of the prayers of men, in that it asks for more than he who utters it is aware of—except he pause and think.

"God's servant and yours has prayed his prayer. Has he paused and taken thought? Is it one prayer? No, it is two—one uttered, the other not. Both have reached the ear of Him Who heareth all supplications, the spoken and the unspoken. Ponder this—keep it in mind. If you would beseech a blessing upon yourself, beware! lest without intent you invoke a curse upon a neighbor at the same time. If you pray for the blessing of rain upon your crop which needs it, by that act you are possibly praying for a curse upon some neighbor's crop which may not need rain and can be injured by it.

"You have heard your servant's prayer—the uttered part of it. I am commissioned of God to put into words the other part of it—that part which the pastor—and also you in your hearts—fervently prayed silently. And ignorantly and unthinkingly? God grant that it was so! You heard these words: 'Grant us victory, O Lord our God!' That is sufficient. The whole of the uttered prayer is compact into those pregnant words. Elaborations were not necessary. When you have prayed for victory you have prayed for many unmentioned results which follow victory—must follow it, cannot help but follow it. Upon the listening spirit of God the Father fell also the unspoken part of the prayer. He commandeth me to put it into words. Listen!

"O Lord our Father, our young patriots, idols of our hearts, go forth to battle—be Thou near them! With them—in spirit—we also go forth from the sweet peace of our beloved firesides to smite the foe. O Lord our God, help us to tear their soldiers to bloody shreds with our shells; help us to cover their smiling fields with the pale forms of their patriot dead; help us to drown the thunder of the guns with shrieks of their wounded, writhing in pain; help us to lay waste their humble homes with hurricanes of fire; help us to wring the hearts of their unoffending widows with unavailing grief; help us to turn them out roofless with their little children to wander unfriended the wastes of their desolated land in rags and hunger and thirst, sports of the sun flames of summer and the icy winds of winter, broken in spirit, worn with travail, imploring Thee for the refuge of the grave and denied it—for our sakes who adore Thee, Lord, blast their hopes, blight their lives, protract their bitter pilgrimage, make heavy their steps, water their way with tears, stain the white snow with the blood of their wounded feet! We ask it, in the spirit of love, of Him Who is the Source of Love, and Who is the ever-faithful refuge and friend of all that are sore beset and seek His aid with humble and contrite hearts. Amen."

[After a pause.] "Ye have prayed it; if ye still desire it, speak! The messenger of the Most High waits".

It was believed afterward that the man was a lunatic, because there was no sense in what he said.

Appendix D

War Is the Health of the State

By Randolph Bourne

From: *Untimely Papers.* New York: B.W. Huebsch, 1919

To most Americans of the classes which consider themselves significant the war [World War I] brought a sense of the sanctity of the State which, if they had had time to think about it, would have seemed a sudden and surprising alteration in their habits of thought. In times of peace, we usually ignore the State in favor of partisan political controversies, or personal struggles for office, or the pursuit of party policies. It is the Government rather than the State with which the politically minded are concerned. The State is reduced to a shadowy emblem which comes to consciousness only on occasions of patriotic holiday.

Government is obviously composed of common and unsanctified men, and is thus a legitimate object of criticism and even contempt. If your own party is in power, things may be assumed to be moving safely enough; but if the opposition is in, then clearly all safety and honor have fled the State. Yet you do not put it to yourself in quite that way. What you think is only that there are rascals to be turned out of a very practical machinery of offices and functions which you take for granted. When we say that Americans are lawless, we usually mean that they are less conscious than other peoples of the august majesty of the institution of the State as it stands behind the objective government of men and laws which we see. In a republic the men who hold office are indistinguishable from the mass. Very few of them possess the slightest personal dignity with which they could endow their political role; even if they ever thought of such a thing. And they have no class distinction to give them glamour. In a republic the Government is obeyed grumblingly, because it has no bedazzlements or sanctities to gild it. If you are a good old-fashioned democrat, you rejoice at this fact, you glory in the plainness of a system where every citizen has become a king. If you are more sophisticated you bemoan the passing of dignity and honor from affairs of State. But in practice, the democrat does not in the least treat his elected citizen with the respect due to a king, nor does the sophisticated citizen pay tribute to the dignity even when he finds it. The republican State has almost no trappings to appeal to the common man's emotions. What it has are of military origin, and in an unmilitary era such as we have passed through since the Civil War, even military trappings have been scarcely seen. In such an era the sense of the State almost fades out of the consciousness of men.

Randolph Silliman Bourne. (Source: photographer unknown, in the public domain, from Wikipedia.)

With the shock of war, however, the State comes into its own again. The Government, with no mandate from the people, without consultation of the people, conducts all the negotiations, the backing and fill-ing, the menaces and explanations, which slowly bring it into collision with some other Government, and gently and irresistibly slides the country into war. For the benefit of proud and haughty citizens, it is fortified with a list of the intolerable insults which have been hurled toward us by the other nations; for the benefit of the liberal and beneficent, it has a convincing set of moral purposes which our going to war will achieve; for the ambitious and aggressive classes, it can gently whisper of a bigger role in the destiny of the world. The result is that, even in those countries where the business of declaring war is theoretically in the hands of representatives of the people, no legislature has ever been known to decline the request of an Executive, which has conducted all foreign affairs in utter privacy and irresponsibility, that it order the nation into battle. Good democrats are wont to feel the crucial difference between a State in which the popular Parliament or Congress declares war, and the State in which an absolute monarch or ruling class declares war. But, put to the stern pragmatic test, the difference is not striking. In the freest of republics as well as in the most tyrannical of empires, all foreign policy, the diplomatic negotiations which produce or forestall war, are equally the private property of the Executive part of the Government, and are equally exposed to no check whatever from popular bodies, or the people voting as a mass themselves.

The moment war is declared, however, the mass of the people, through some spiritual alchemy, become convinced that they have willed and executed the deed themselves. They then, with the exception of a few malcontents, proceed to allow themselves to be regimented, coerced, deranged in all the environments of their lives, and turned into a solid manufactory of destruction toward whatever other people may have, in the appointed scheme of things, come within the range of the Government's disapprobation. The citizen throws off his contempt and indifference to Government, identifies himself with its purposes, revives all his military memories and symbols, and the State once more walks, an august presence, through the imaginations of men. Patriotism becomes the dominant feeling, and produces immediately that intense and hopeless confusion between the relations which the individual bears and should bear toward the society of which he is a part.

The patriot loses all sense of the distinction between State, nation, and government. In our quieter moments, the Nation or Country forms the basic idea of society. We think vaguely of a loose population spreading over a certain geographical portion of the earth's surface, speaking a common language, and living in a homogeneous civilization. Our idea of Country concerns itself with the non-political aspects of a people, its ways of living, its personal traits, its literature and art, its characteristic

attitudes toward life. We are Americans because we live in a certain
bounded territory, because our ancestors have carried on a great enterprise of pioneering and colonization, because we live in certain kinds of
communities which have a certain look and express their aspirations in
certain ways. We can see that our civilization is different from contigu-
ous civilizations like the Indian and Mexican. The institutions of our
country form a certain network which affects us vitally and intrigues our
thoughts in a way that these other civilizations do not. We are a part of
Country, for better or for worse. We have arrived in it through the operation of physiological laws, and not in any way through our own
choice. By the time we have reached what are called years of discretion,
its influences have molded our habits, our values, our ways of thinking,
so that however aware we may become, we never really lose the stamp of
our civilization, or could be mistaken for the child of any other country.
Our feeling for our fellow countrymen is one of similarity or of mere
acquaintance. We may be intensely proud of and congenial to our particular network of civilization, or we may detest most of its qualities and
rage at its defects. This does not alter the fact that we are inextricably
bound up in it. The Country, as an inescapable group into which we
are born, and which makes us its particular kind of a citizen of the
world, seems to be a fundamental fact of our consciousness, an irre-
ducible minimum of social feeling.

Now this feeling for country is essentially noncompetitive; we think
of our own people merely as living on the earth's surface along with
other groups, pleasant or objectionable as they may be, but fundamentally as sharing the earth with them. In our simple conception of country there is no more feeling of rivalry with other peoples than there is in
our feeling for our family. Our interest turns within rather than without, is intensive and not belligerent. We grow up and our imaginations
gradually stake out the world we live in, they need no greater conscious
satisfaction for their gregarious impulses than this sense of a great mass
of people to whom we are more or less attuned, and in whose institutions we are functioning. The feeling for country would be an uninflat-
able maximum were it not for the ideas of State and Government which
are associated with it. Country is a concept of peace, of tolerance, of living and letting live. But State is essentially a concept of power, of competition: it signifies a group in its aggressive aspects. And we have the
misfortune of being born not only into a country but into a State, and
as we grow up we learn to mingle the two feelings into a hopeless confusion.

The State is the country acting as a political unit, it is the group acting as a repository of force, determiner of law, arbiter of justice. Interna-

tional politics is a "power politics" because it is a relation of States and that is what States infallibly and calamitously are, huge aggregations of human and industrial force that may be hurled against each other in war. When a country acts as a whole in relation to another country, or in imposing laws on its own inhabitants, or in coercing or punishing individuals or minorities, it is acting as a State. The history of America as a country is quite different from that of America as a State. In one case it is the drama of the pioneering conquest of the land, of the growth of wealth and the ways in which it was used, of the enterprise of education, and the carrying out of spiritual ideals, of the struggle of economic classes. But as a State, its history is that of playing a part in the world, making war, obstructing international trade, preventing itself from being split to pieces, punishing those citizens whom society agrees are offensive, and collecting money to pay for all.

Government on the other hand is synonymous with neither State nor Nation. It is the machinery by which the nation, organized as a State, carries out its State functions. Government is a framework of the administration of laws, and the carrying out of the public force. Government is the idea of the State put into practical operation in the hands of definite, concrete, fallible men. It is the visible sign of the invisible grace. It is the word made flesh. And it has necessarily the limitations inherent in all practicality. Government is the only form in which we can envisage the State, but it is by no means identical with it. That the State is a mystical conception is something that must never be forgotten. Its glamour and its significance linger behind the framework of Government and direct its activities.

Wartime brings the ideal of the State out into very clear relief, and reveals attitudes and tendencies that were hidden. In times of peace the sense of the State flags in a republic that is not militarized. For war is essentially the health of the State. The ideal of the State is that within its territory its power and influence should be universal. As the Church is the medium for the spiritual salvation of man, so the State is thought of as the medium for his political salvation. Its idealism is a rich blood flowing to all the members of the body politic. And it is precisely in war that the urgency for union seems greatest, and the necessity for universality seems most unquestioned. The State is the organization of the herd to act offensively or defensively against another herd similarly organized. The more terrifying the occasion for defense, the closer will become the organization and the more coercive the influence upon each member of the herd. War sends the current of purpose and activity flowing down to the lowest level of the herd, and to its most remote branches. All the activities of society are linked together as fast as possible to this central purpose of making a military offensive or a military defense, and the State becomes what in peacetimes it has vainly struggled to become—the inexorable arbiter and determinant of men's business and attitudes and opinions. The slack is taken up, the cross-currents fade out, and the nation moves lumberingly and slowly, but with ever accelerated speed and integration, toward the great end, toward the "peacefulness of being at war," of which L.P. Jacks has so unforgettably spoken.

The classes which are able to play an active and not merely a passive role in the organization for war get a tremendous liberation of activity and energy. Individuals are jolted out of their old routine, many of them are given new positions of responsibility, new techniques must be learned. Wearing home ties are broken and women who would have remained attached with infantile bonds are liberated for service overseas. A vast sense of rejuvenescence pervades the significant classes, a sense of new importance in the world. Old national ideals are taken out, re-adapted to the purpose and used as universal touchstones, or molds into which all thought is poured. Every individual citizen who in peacetimes had no function to perform by which he could imagine himself an expression or living fragment of the State becomes an active amateur agent of the Government in reporting spies and disloyalists, in raising Government funds, or in propagating such measures as are considered necessary by officialdom. Minority opinion, which in times of peace, was only irritating and could not be dealt with by law unless it was conjoined with actual crime, becomes, with the outbreak of war, a case for outlawry. Criticism of the State, objections to war, lukewarm opinions concerning the necessity or the beauty of conscription, are made subject to ferocious penalties, far exceeding in severity those affixed to actual pragmatic crimes. Public opinion, as expressed in the newspapers, and the pulpits and the schools, becomes one solid block. "Loyalty," or rather war orthodoxy, becomes the sole test for all professions, techniques, occupations. Particularly is this true in the sphere of the intellectual life. There the smallest taint is held to spread over the whole soul, so that a professor of physics is ipso facto disqualified to teach physics or to hold honorable place in a university—the republic of learning—if he is at all unsound on the war. Even mere association with persons thus tainted is considered to disqualify a teacher. Anything pertaining to the enemy becomes taboo. His books are suppressed wherever possible, his language is forbidden. His artistic products are considered to convey in the subtlest spiritual way taints of vast poison to the soul that permits itself to enjoy them. So enemy music is suppressed, and energetic measures of opprobrium taken against those whose artistic consciences are not ready to perform such an act of self-sacrifice. The rage for loyal conformity works impartially, and often in diametric opposition to other orthodoxies and traditional conformities, or even ideals. The triumphant orthodoxy of the State is shown at its apex perhaps when Christian preachers lose their pulpits for taking in more or less literal terms the Sermon on the Mount, and Christian zealots are sent to prison for twenty years for distributing tracts which argue that war is unscriptural.

War is the health of the State. It automatically sets in motion throughout society those irresistible forces for uniformity, for passionate cooperation with the Government in coercing into obedience the minority groups and individuals which lack the larger herd sense. The machinery of government sets and enforces the drastic penalties; the minorities are either intimidated into silence, or brought slowly around by a subtle process of persuasion which may seem to them really to be converting them. Of course, the ideal of perfect loyalty, perfect uniformity is never really attained. The classes upon whom the amateur work of coercion falls are unwearied in their zeal, but often their agitation instead of converting, merely serves to stiffen their resistance. Minorities are rendered sullen, and some intellectual opinion bitter and satirical. But in general, the nation in wartime attains a uniformity of feeling, a hierarchy of values culminating at the undisputed apex of the State ideal, which could not possibly be produced through any other agency than war. Loyalty—or mystic devotion to the State—becomes the major imagined human value. Other values, such as artistic creation, knowledge, reason, beauty, the enhancement of life, are instantly and almost unanimously sacrificed, and the significant classes who have constituted themselves the amateur agents of the State are engaged not only in sacrificing these values for themselves but in coercing all other persons into sacrificing them.

War—or at least modern war waged by a democratic republic against a powerful enemy—seems to achieve for a nation almost all that the most inflamed political idealist could desire. Citizens are no longer indifferent to their Government, but each cell of the body politic is brimming with life and activity. We are at last on the way to full realization of that collective community in which each individual somehow contains the virtue of the whole. In a nation at war, every citizen identifies himself with the whole, and feels immensely strengthened in that identification. The purpose and desire of the collective community live in each person who throws himself wholeheartedly into the cause of war. The impeding distinction between society and the individual is almost blotted out. At war, the individual becomes almost identical with his society. He achieves a superb self-assurance, an intuition of the rightness of all his ideas and emotions, so that in the suppression of opponents or heretics he is invincibly strong; he feels behind him all the power of the collective community. The individual as social being in war seems to have achieved almost his apotheosis. Not for any religious impulse could the American nation have been expected to show such devotion en masse, such sacrifice and labor. Certainly not for any secular good, such as universal education or the subjugation of nature, would it have poured forth its treasure and its life, or would it have permitted such stern coercive measures to be taken against it, such as conscripting its money and its men. But for the sake of a war of offensive self-defense, undertaken to support a difficult cause to the slogan of "democracy," it would reach the highest level ever known of collective effort.

For these secular goods, connected with the enhancement of life, the education of man and the use of the intelligence to realize reason and beauty in the nation's communal living, are alien to our traditional ideal of the State. The State is intimately connected with war, for it is the organization of the collective community when it acts in a political manner, and to act in a political manner towards a rival group has meant, throughout all history—war.

There is nothing invidious in the use of the term "herd" in connection with the State. It is merely an attempt to reduce closer to first principles the nature of this institution in the shadow of which we all live, move, and have our being. Ethnologists are generally agreed that human society made its first appearance as the human pack and not as a collection of individuals or of couples. The herd is in fact the original unit, and only as it was differentiated did personal individuality develop. All the most primitive surviving tribes of men are shown to live in a very complex but very rigid social organization where opportunity for individuation is scarcely given. These tribes remain strictly organized herds, and the difference between them and the modern State is one of degree of sophistication and variety of organization, and not of kind.

Psychologists recognize the gregarious impulse as one of the strongest primitive pulls which keeps together the herds of the different species of higher animals. Mankind is no exception. Our pugnacious evolutionary history has prevented the impulse from ever dying out. This gregarious impulse is the tendency to imitate, to conform, to coalesce together, and is most powerful when the herd believes itself threatened with attack. Animals crowd together for protection, and men become most conscious of their collectivity at the threat of war. Consciousness of collectivity brings confidence and a feeling of massed strength, which in turn arouses pugnacity and the battle is on. In civilized man, the gregarious impulse acts not only to produce concerted action for defense, but also to produce identity of opinion. Since thought is a form of behavior, the gregarious impulse floods up into its realms and demands that sense of uniform thought which wartime produces so successfully. And it is in this flooding of the conscious life of society that gregariousness works its havoc.

For just as in modern societies the sex instinct is enormously oversupplied for the requirements of human propagation, so the gregarious impulse is enormously oversupplied for the work of protection which it is called upon to perform. It would be quite enough if we were gregarious enough to enjoy the companionship of others, to be able to cooperate with them, and to feel a slight malaise at solitude. Unfortunately, however, this impulse is not content with these reasonable and healthful demands, but insists that like-mindedness shall prevail everywhere, in all departments of life. So that all human progress, all novelty, and nonconformity, must be carried against the resistance of this tyrannical herd instinct which drives the individual into obedience and conformity with the majority. Even in the most modern and enlightened societies this impulse shows little sign of abating. As it is driven by inexorable economic demand out of the sphere of utility, it seems to fasten itself ever more fiercely in the realm of feeling and opinion, so that conformity comes to be a thing aggressively desired and demanded.

The gregarious impulse keeps its hold all the more virulently because when the group is in motion or is taking any positive action, this feeling of being with and supported by the collective herd very greatly feeds that will to power, the nourishment of which the individual organism so constantly demands. You feel powerful by conforming, and you feel forlorn and helpless if you are out of the crowd. While even if you do not get any access of power by thinking and feeling just as everybody else in your group does, you get at least the warm feeling of obedience, the soothing irresponsibility of protection.

Joining as it does to these very vigorous tendencies of the individual—the pleasure in power and the pleasure in obedience—this gregarious impulse becomes irresistible in society. War stimulates it to the highest possible degree, sending the influences of its mysterious herd-current with its inflations of power and obedience to the farthest reaches of the society, to every individual and little group that can possibly be affected. And it is these impulses which the State—the organization of the entire herd, the entire collectivity—is founded on and makes use of.

There is, of course, in the feeling toward the State a large element of pure filial mysticism. The sense of insecurity, the desire for protection, sends one's desire back to the father and mother, with whom is associated the earliest feelings of protection. It is not for nothing that one's State is still thought of as Father or Motherland, that one's relation toward it is conceived in terms of family affection. The war has shown that nowhere under the shock of danger have these primitive childlike attitudes failed to assert themselves again, as much in this country as anywhere. If we have not the intense Father-sense of the German who worships his Vaterland, at least in Uncle Sam we have a symbol of protecting, kindly authority, and in the many Mother-posters of the Red Cross, we see how easily in the more tender functions of war service, the ruling organization is conceived in family terms. A people at war have become in the most literal sense obedient, respectful, trustful children again, full of that naïve faith in the all-wisdom and all-power of the adult who takes care of them, imposes his mild but necessary rule upon them and in whom they lose their responsibility and anxieties. In this recrudescence of the child, there is great comfort, and a certain influx of power. On most people the strain of being an independent adult weighs heavily, and upon none more than those members of the significant classes who have had bequeathed to them or have assumed the responsibilities of governing. The State provides the convenientest of symbols under which these classes can retain all the actual pragmatic satisfaction of governing, but can rid themselves of the psychic burden of adulthood. They continue to direct industry and government and all the institutions of society pretty much as before, but in their own conscious eyes and in the eyes of the general public, they are turned from their selfish and predatory ways, and have become loyal servants of society, or something greater than they—the State. The man who moves from the direction of a large business in New York to a post in the war management industrial service in Washington does not apparently alter very much his power or his administrative technique. But psychically, what a transfiguration has occurred! His is now not only the power but the glory! And his sense of satisfaction is directly proportional not to the genuine amount of personal sacrifice that may be involved in the change but to the extent to which he retains his industrial prerogatives and sense of command.

From members of this class a certain insuperable indignation arises if the change from private enterprise to State service involves any real loss of power and personal privilege. If there is to be pragmatic sacrifice, let it be, they feel, on the field of honor, in the traditionally acclaimed deaths by battle, in that detour to suicide, as Nietzsche calls war. The

State in wartime supplies satisfaction for this very real craving, but its chief value is the opportunity it gives for this regression to infantile attitudes. In your reaction to an imagined attack on your country or an insult to its government, you draw closer to the herd for protection, you conform in word and deed, and you insist vehemently that everybody else shall think, speak, and act together. And you fix your adoring gaze upon the State, with a truly filial look, as upon the Father of the flock, the quasi-personal symbol of the strength of the herd, and the leader and determinant of your definite action and ideas.

The members of the working classes, that portion at least which does not identify itself with the significant classes and seek to imitate it and rise to it, are notoriously less affected by the symbolism of the State, or, in other words, are less patriotic than the significant classes. For theirs is neither the power nor the glory. The State in wartime does not offer them the opportunity to regress, for, never having acquired social adulthood, they cannot lose it. If they have been drilled and regimented, as by the industrial regime of the last century, they go out docilely enough to do battle for their State, but they are almost entirely without that filial sense and even without that herd-intellect sense which operates so powerfully among their "betters." They live habitually in an industrial serfdom, by which, though nominally free, they are in practice as a class bound to a system of machine-production the implements of which they do not own, and in the distribution of whose product they have not the slightest voice, except what they can occasionally exert by a veiled intimidation which draws slightly more of the product in their direction. From such serfdom, military conscription is not so great a change. But into the military enterprise they go, not with those hurrahs of the significant classes whose instincts war so powerfully feeds, but with the same apathy with which they enter and continue in the industrial enterprise.

From this point of view, war can be called almost an upper-class sport. The novel interests and excitements it provides, the inflations of power, the satisfaction it gives to those very tenacious human impulses—gregariousness and parent-regression—endow it with all the qualities of a luxurious collective game which is felt intensely just in proportion to the sense of significant rule the person has in the class division of his society. A country at war—particularly our own country at war—does not act as a purely homogeneous herd. The significant classes have all the herd-feeling in all its primitive intensity, but there are barriers, or at least differentials of intensity, so that this feeling does not flow freely without impediment throughout the entire nation. A modern country represents a long historical and social process of disaggregation of the herd. The nation at peace is not a group, it is a network of myriads of groups representing the cooperation and similar feeling of men on all sorts of planes and in all sorts of human interests and enterprises. In every modern industrial country, there are parallel planes of economic classes with divergent attitudes and institutions and interests—bourgeois and proletariat, with their many subdivisions according to power and function, and even their interweaving, such as those more highly skilled workers who habitually identify themselves with the owning and the significant classes and strive to raise themselves to the bourgeois level, imitating their cultural standards and manners. Then there are religious groups with a certain definite, though weakening sense of kinship, and there are the powerful ethnic groups which behave almost as cultural colonies in the New World, clinging tenaciously to language and historical tradition, though their herdishness is usually founded on cultural rather than State symbols. There are even certain vague sectional groupings. All these small sects, political parties, classes, levels, interests, may act as foci for herd-feelings. They intersect and interweave, and the same person may be a member of several different groups lying at different planes. Different occasions will set off his herd-feeling in one direction or another. In a religious crisis he will be intensely conscious of the necessity that his sect (or sub-herd) may prevail, in a political campaign, that his party shall triumph.

To the spread of herd-feeling, therefore, all these smaller herds offer resistance. To the spread of that herd-feeling which arises from the threat of war, and which would normally involve the entire nation, the only groups which make serious resistance are those, of course, which continue to identify themselves with the other nation from which they or their parents have come. In times of peace they are for all practical purposes citizens of their new country. They keep alive their ethnic traditions more as a luxury than anything. Indeed these traditions tend rapidly to die out except where they connect with some still unresolved nationalistic cause abroad, with some struggle for freedom, or some irredentism. If they are consciously opposed by a too invidious policy of Americanism, they tend to be strengthened. And in time of war, these ethnic elements which have any traditional connection with the enemy, even though most of the individuals may have little real sympathy with the enemy's cause, are naturally lukewarm to the herd-feeling of the nation which goes back to State traditions in which they have no share. But to the natives imbued with State-feeling, any such resistance or apathy is intolerable. This herd-feeling, this newly awakened consciousness of the State, demands universality. The leaders of the significant classes, who feel most intensely this State compulsion, demand a 100 percent

Americanism, among 100 percent of the population. The State is a jealous God and will brook no rivals. Its sovereignty must pervade every one, and all feeling must be run into the stereotyped forms of romantic patriotic militarism which is the traditional expression of the State herd-feeling.

Thus arises conflict within the State. War becomes almost a sport between the hunters and the hunted. The pursuit of enemies within outweighs in psychic attractiveness the assault on the enemy without. The whole terrific force of the State is brought to bear against the heretics. The nation boils with a slow insistent fever. A white terrorism is carried on by the Government against pacifists, socialists, enemy aliens, and a milder unofficial persecution against all persons or movements that can be imagined as connected with the enemy. War, which should be the health of the State, unifies all the bourgeois elements and the common people, and outlaws the rest. The revolutionary proletariat shows more resistance to this unification, is, as we have seen, psychically out of the current. Its vanguard, as the I.W.W., is remorselessly pursued, in spite of the proof that it is a symptom, not a cause, and its persecution increases the disaffection of labor and intensifies the friction instead of lessening it.

But the emotions that play around the defense of the State do not take into consideration the pragmatic results. A nation at war, led by its significant classes, is engaged in liberating certain of its impulses which have had all too little exercise in the past. It is getting certain satisfactions, and the actual conduct of the war or the condition of the country are really incidental to the enjoyment of new forms of virtue and power and aggressiveness. If it could be shown conclusively that the persecution of slightly disaffected elements actually increased enormously the difficulties of production and the organization of the war technique, it would be found that public policy would scarcely change. The significant classes must have their pleasure in hunting down and chastising everything that they feel instinctively to be not imbued with the current State enthusiasm, though the State itself be actually impeded in its efforts to carry out those objects for which they are passionately contending. The best proof of this is that with a pursuit of plotters that has continued with ceaseless vigilance ever since the beginning of the war in Europe, the concrete crimes unearthed and punished have been fewer than those prosecutions for the mere crime of opinion or the expression of sentiments critical of the State or the national policy. The punishment for opinion has been far more ferocious and unintermittent than the punishment of pragmatic crime. Unimpeachable Anglo-Saxon Americans who were freer of pacifist or socialist utterance than the

State-obsessed ruling public opinion, received heavier penalties and even greater opprobrium, in many instances, than the definitely hostile German plotter. A public opinion which, almost without protest, accepts as just, adequate, beautiful, deserved, and in fitting harmony with ideals of liberty and freedom of speech, a sentence of twenty years in prison for mere utterances, no matter what they may be, shows itself to be suffering from a kind of social derangement of values, a sort of social neurosis, that deserves analysis and comprehension.

On our entrance into the war, there were many persons who predicted exactly this derangement of values, who feared lest democracy suffer more at home from an America at war than could be gained for democracy abroad. That fear has been amply justified. The question whether the American nation would act like an enlightened democracy going to war for the sake of high ideals, or like a State-obsessed herd, has been decisively answered. The record is written and cannot be erased. History will decide whether the terrorization of opinion and the regimentation of life were justified under the most idealistic of democratic administrations. It will see that when the American nation had ostensibly a chance to conduct a gallant war, with scrupulous regard to the safety of democratic values at home, it chose rather to adopt all the most obnoxious and coercive techniques of the enemy and of the other countries at war, and to rival in intimidation and ferocity of punishment the worst governmental systems of the age. For its former unconsciousness and disrespect of the State ideal, the nation apparently paid the penalty in a violent swing to the other extreme. It acted so exactly like a herd in its irrational coercion of minorities that there is no artificiality in interpreting the progress of the war in terms of the herd psychology. It unwittingly brought out into the strongest relief the true characteristics of the State and its intimate alliance with war. It provided for the enemies of war and the critics of the State the most telling arguments possible. The new passion for the State ideal unwittingly set in motion and encouraged forces that threaten very materially to reform the State. It has shown those who are really determined to end war that the problem is not the mere simple one of finishing a war that will end war.

For war is a complicated way in which a nation acts, and it acts so out of a spiritual compulsion which pushes it on, perhaps against all its interests, all its real desires, and all its real sense of values. It is States that make wars and not nations, and the very thought and almost necessity of war is bound up with the ideal of the State. Not for centuries have nations made war; in fact the only historical example of nations making war is the great barbarian invasions into southern Europe, the invasions of Russia from the East, and perhaps the sweep of Islam through northern Africa into Europe after Mohammed's death. And the motivations for such wars were either the restless expansion of migratory tribes or the flame of religious fanaticism. Perhaps these great movements could scarcely be called wars at all, for war implies an organized people drilled and led: in fact, it necessitates the State. Ever since Europe has had any such organization, such huge conflicts between nations—nations, that is, as cultural groups—have been unthinkable. It is preposterous to assume that for centuries in Europe there would have been any possibility of a people en masse (with their own leaders, and not with the leaders of their duly constituted State) rising up and overflowing their borders in a war raid upon a neighboring people. The wars of the Revolutionary armies of France were clearly in defense of an imperiled freedom, and, moreover, they were clearly directed not against other peoples, but against the autocratic governments that were combining to crush the Revolution. There is no instance in history of a genuinely national war. There are instances of national defenses, among primitive civilizations such as the Balkan peoples, against intolerable invasion by neighboring despots or oppression. But war, as such, cannot occur except in a system of competing States, which have relations with each other through the channels of diplomacy.

War is a function of this system of States, and could not occur except in such a system. Nations organized for internal administration, nations organized as a federation of free communities, nations organized in any way except that of a political centralization of a dynasty, or the reformed descendant of a dynasty, could not possibly make war upon each other. They would not only have no motive for conflict, but they would be unable to muster the concentrated force to make war effective. There might be all sorts of amateur marauding, there might be guerrilla expeditions of group against group, but there could not be that terrible war en masse of the national State, that exploitation of the nation in the interests of the State, that abuse of the national life and resource in the frenzied mutual suicide, which is modern war.

It cannot be too firmly realized that war is a function of States and not of nations, indeed that it is the chief function of States. War is a very artificial thing. It is not the naïve spontaneous outburst of herd pugnacity; it is no more primary than is formal religion. War cannot exist without a military establishment, and a military establishment cannot exist without a State organization. War has an immemorial tradition and heredity only because the State has a long tradition and heredity. But they are inseparably and functionally joined. We cannot crusade against war without crusading implicitly against the State. And we cannot expect, or take measures to ensure, that this war is a war to end war, unless at the same time we take measures to end the State in its traditional form. The State is not the nation, and the State can be modified and even abolished in its present form, without harming the nation. On the contrary, with the passing of the dominance of the State, the genuine life-enhancing forces of the nation will be liberated. If the State's chief function is war, then the State must suck out of the nation a large part of its energy for its purely sterile purposes of defense and aggression. It devotes to waste or to actual destruction as much as it can of the vitality of the nation. No one will deny that war is a vast complex of life-destroying and life-crippling forces. If the State's chief function is war, then it is chiefly concerned with coordinating and developing the powers and techniques which make for destruction. And this means not only the actual and potential destruction of the enemy, but of the nation at home as well. For the very existence of a State in a system of States means that the nation lies always under a risk of war and invasion, and the calling away of energy into military pursuits means a crippling of the productive and life-enhancing processes of the national life.

All this organization of death-dealing energy and technique is not a natural but a very sophisticated process. Particularly in modern nations, but also all through the course of modern European history, it could never exist without the State. For it meets the demands of no other institution, it follows the desires of no religious, industrial, political group. If the demand for military organization and a military establishment seems to come not from the officers of the State but from the public, it is only that it comes from the State-obsessed portion of the public, those groups which feel most keenly the State ideal. And in this country we have had evidence all too indubitable how powerless the pacifically minded officers of State may be in the face of a State obsession of the significant classes. If a powerful section of the significant classes feels more intensely the attitudes of the State, then they will most infallibly mold the Government in time to their wishes, bring it back to act as the embodiment of the State which it pretends to be. In every country we have seen groups that were more loyal than the king— more patriotic than the Government—the Ulsterites in Great Britain, the Junkers in Prussia, l'Action Française in France, our patrioteers in America. These groups exist to keep the steering wheel of the State straight, and they prevent the nation from ever veering very far from the State ideal.

Militarism expresses the desires and satisfies the major impulse only of this class. The other classes, left to themselves, have too many necessities and interests and ambitions, to concern themselves with so expensive and destructive a game. But the State-obsessed group is either able to get control of the machinery of the State or to intimidate those in control, so that it is able through use of the collective force to regiment the other grudging and reluctant classes into a military program. State idealism percolates down through the strata of society; capturing groups and individuals just in proportion to the prestige of this dominant class. So that we have the herd actually strung along between two extremes, the militaristic patriots at one end, who are scarcely distinguishable in attitude and animus from the most reactionary Bourbons of an Empire, and unskilled labor groups, which entirely lack the State sense. But the State acts as a whole, and the class that controls governmental machinery can swing the effective action of the herd as a whole. The herd is not actually a whole, emotionally. But by an ingenious mixture of cajolery, agitation, intimidation, the herd is licked into shape, into an effective mechanical unity, if not into a spiritual whole. Men are told simultaneously that they will enter the military establishment of their own volition, as their splendid sacrifice for their country's welfare, and that if they do not enter they will be hunted down and punished with the most horrid penalties; and under a most indescribable confusion of democratic pride and personal fear they submit to the destruction of their livelihood if not their lives, in a way that would formerly have seemed to them so obnoxious as to be incredible.

In this great herd machinery, dissent is like sand in the bearings. The State ideal is primarily a sort of blind animal push toward military unity. Any difference with that unity turns the whole vast impulse toward crushing it. Dissent is speedily outlawed, and the Government, backed by the significant classes and those who in every locality, however small, identify themselves with them, proceeds against the outlaws, regardless of their value to the other institutions of the nation, or to the effect their persecution may have on public opinion. The herd becomes divided into the hunters and the hunted, and war enterprise becomes not only a technical game but a sport as well.

It must never be forgotten that nations do not declare war on each other, nor in the strictest sense is it nations that fight each other. Much has been said to the effect that modern wars are wars of whole peoples and not of dynasties. Because the entire nation is regimented and the whole resources of the country are levied on for war, this does not mean that it is the country qua country which is fighting. It is the country organized as a State that is fighting, and only as a State would it possibly fight. So literally it is States which make war on each other and not peoples. Governments are the agents of States, and it is Governments which declare war on each other, acting truest to form in the interests of the great State ideal they represent. There is no case known in modern times of the people being consulted in the initiation of a war. The present demand for "democratic control" of foreign policy indicates how completely, even in the most democratic of modern nations, foreign policy has been the secret private possession of the executive branch of the Government.

However representative of the people Parliaments and Congresses may be in all that concerns the internal administration of a country's political affairs, in international relations it has never been possible to maintain that the popular body acted except as a wholly mechanical ratifier of the Executive's will. The formality by which Parliaments and Congresses declare war is the merest technicality. Before such a declaration can take place, the country will have been brought to the very brink of war by the foreign policy of the Executive. A long series of steps on the downward path, each one more fatally committing the unsuspecting country to a warlike course of action, will have been taken without either the people or its representatives being consulted or expressing its feeling. When the declaration of war is finally demanded by the Executive, the Parliament or Congress could not refuse it without reversing the course of history, without repudiating what has been representing itself in the eyes of the other States as the symbol and interpreter of the nation's will and animus. To repudiate an Executive at that time would be to publish to the entire world the evidence that the country had been grossly deceived by its own Government, that the country with an almost criminal carelessness had allowed its Government to commit it to gigantic national enterprises in which it had no heart. In such a crisis, even a Parliament which in the most democratic States represents the common man and not the significant classes who most strongly cherish the State ideal, will cheerfully sustain the foreign policy which it understands even less than it would care for if it understood, and will vote almost unanimously for an incalculable war, in which the nation may be brought well nigh to ruin. That is why the referendum which was advocated by some people as a test of American sentiment in entering the war was considered even by thoughtful democrats to be something subtly improper. The die had been cast. Popular whim could only derange and bungle monstrously the majestic march of State policy in its new crusade for the peace of the world. The irresistible State ideal got hold of the bowels of men. Whereas up to this time, it had been irreproachable to be neutral in word and deed, for the foreign policy of the State had so decided it, henceforth it became the most arrant crime to remain neutral. The Middle West, which had been soddenly pacifistic in our days of neutrality, became in a few months just as soddenly bellicose, and in its zeal for witch-burnings and its scent for enemies within gave precedence to no section of the country. The herd-mind followed faithfully the State-mind and, the agitation for a referendum being soon forgotten, the country fell into the universal conclusion that, since its Congress had formally declared the war, the nation itself had in the most solemn and universal way devised and brought on the entire affair. Oppression of minorities became justified on the plea that the latter were perversely resisting the rationally constructed and solemnly declared will of a majority of the nation. The herd coalescence of opinion which became inevitable the moment the State had set flowing the war attitudes became interpreted as a prewar popular decision, and disinclination to bow to the herd was treated as a monstrously antisocial act. So that the State, which had vigorously resisted the idea of a referendum and clung tenaciously and, of course, with entire success to its autocratic and absolute control of foreign policy, had the pleasure of seeing the country, within a few months, given over to the retrospective impression that a genuine referendum had taken place. When once a country has lapped up these State attitudes, its memory fades; it conceives itself not as merely accepting, but of having itself willed, the whole policy and technique of war. The significant classes, with their trailing satellites, identify themselves with the State, so that what the State, through the agency of the Government, has willed, this majority conceives itself to have willed.

All of which goes to show that the State represents all the autocratic, arbitrary, coercive, belligerent forces within a social group, it is a sort of complexus of everything most distasteful to the modern free creative spirit, the feeling for life, liberty, and the pursuit of happiness. War is the health of the State. Only when the State is at war does the modern society function with that unity of sentiment, simple uncritical patriotic devotion, cooperation of services, which have always been the ideal of the State lover. With the ravages of democratic ideas, however, the modern republic cannot go to war under the old conceptions of autocracy and death-dealing belligerency. If a successful animus for war requires a renaissance of State ideals, they can only come back under democratic forms, under this retrospective conviction of democratic control of foreign policy, democratic desire for war, and particularly of this identification of the democracy with the State. How unregenerate the ancient State may be, however, is indicated by the laws against sedition, and by the Government's unreformed attitude on foreign policy. One of the first demands of the more farseeing democrats in the democracies of the Alliance was that secret diplomacy must go. The war

was seen to have been made possible by a web of secret agreements between States, alliances that were made by Governments without the
shadow of popular support or even popular knowledge, and vague, half-
understood commitments that scarcely reached the stage of a treaty or
agreement, but which proved binding in the event. Certainly, said these
democratic thinkers, war can scarcely be avoided unless this poisonous
underground system of secret diplomacy is destroyed, this system by
which a nation's power, wealth, and manhood may be signed away like
a blank check to an allied nation to be cashed in at some future crisis.
Agreements which are to affect the lives of whole peoples must be made
between peoples and not by Governments, or at least by their representatives in the full glare of publicity and criticism.

Such a demand for "democratic control of foreign policy" seemed
axiomatic. Even if the country had been swung into war by steps taken
secretly and announced to the public only after they had been con-
summated, it was felt that the attitude of the American State toward
foreign policy was only a relic of the bad old days and must be superseded in the new order. The American President himself, the liberal
hope of the world, had demanded, in the eyes of the world, open diplomacy, agreements freely and openly arrived at. Did this mean a genuine transference of power in this most crucial of State functions from
Government to people? Not at all. When the question recently came to
a challenge in Congress, and the implications of open discussion were
somewhat specifically discussed, and the desirabilities frankly com-
mended, the President let his disapproval be known in no uncertain
way. No one ever accused Mr. Wilson of not being a State idealist, and
whenever democratic aspirations swung ideals too far out of the State
orbit, he could be counted on to react vigorously. Here was a clear case
of conflict between democratic idealism and the very crux of the concept of the State. However unthinkingly he might have been led on to
encourage open diplomacy in his liberalizing program, when its implica-
tion was made vivid to him, he betrayed how mere a tool the idea had
been in his mind to accentuate America's redeeming role. Not in any
sense as a serious pragmatic technique had he thought of a genuinely
open diplomacy. And how could he? For the last stronghold of State
power is foreign policy. It is in foreign policy that the State acts most
concentratedly as the organized herd, acts with fullest sense of aggres-
sive-power, acts with freest arbitrariness. In foreign policy, the State is
most itself. States, with reference to each other, may be said to be in a
continual state of latent war. The "armed truce," a phrase so familiar
before 1914, was an accurate description of the normal relation of
States when they are not at war. Indeed, it is not too much to say that

the normal relation of States is war. Diplomacy is a disguised war, in which States seek to gain by barter and intrigue, by the cleverness of wits, the objectives which they would have to gain more clumsily by means of war. Diplomacy is used while the States are recuperating from conflicts in which they have exhausted themselves. It is the wheedling and the bargaining of the worn-out bullies as they rise from the ground and slowly restore their strength to begin fighting again. If diplomacy had been a moral equivalent for war, a higher stage in human progress, an inestimable means of making words prevail instead of blows, militarism would have broken down and given place to it. But since it is a mere temporary substitute, a mere appearance of war's energy under another form, a surrogate effect is almost exactly proportioned to the armed force behind it. When it fails, the recourse is immediate to the military technique whose thinly veiled arm it has been. A diplomacy that was the agency of popular democratic forces in their non-State manifestations would be no diplomacy at all. It would be no better than the Railway or Education commissions that are sent from one country to another with rational constructive purpose. The State, acting as a diplomatic-military ideal, is eternally at war. Just as it must act arbitrarily and autocratically in time of war, it must act in time of peace in this particular role where it acts as a unit. Unified control is necessarily autocratic control. Democratic control of foreign policy is therefore a contradiction in terms. Open discussion destroys swiftness and certainty of action. The giant State is paralyzed. Mr. Wilson retains his full ideal of the State at the same time that he desires to eliminate war. He wishes to make the world safe for democracy as well as safe for diplomacy. When the two are in conflict, his clear political insight, his idealism of the State, tells him that it is the naïver democratic values that must be sacrificed. The world must primarily be made safe for diplomacy. The State must not be diminished.

What is the State essentially? The more closely we examine it, the more mystical and personal it becomes. On the Nation we can put our hand as a definite social group, with attitudes and qualities exact enough to mean something. On the Government we can put our hand as a certain organization of ruling functions, the machinery of lawmaking and law-enforcing. The Administration is a recognizable group of political functionaries, temporarily in charge of the government. But the State stands as an idea behind them all, eternal, sanctified, and from it Government and Administration conceive themselves to have the breath of life. Even the nation, especially in times of war—or at least, its significant classes—considers that it derives its authority and its purpose from the idea of the State. Nation and State are scarcely differentiated, and the concrete, practical, apparent facts are sunk in the symbol. We reverence not our country but the flag. We may criticize ever so severely our country, but we are disrespectful to the flag at our peril. It is the flag and the uniform that make men's heart beat high and fill them with noble emotions, not the thought of and pious hopes for America as a free and enlightened nation.

It cannot be said that the object of emotion is the same, because the flag is the symbol of the nation, so that in reverencing the American flag we are reverencing the nation. For the flag is not a symbol of the country as a cultural group, following certain ideals of life, but solely a symbol of the political State, inseparable from its prestige and expansion. The flag is most intimately connected with military achievement, military memory. It represents the country not in its intensive life, but in its far-flung challenge to the world. The flag is primarily the banner of war; it is allied with patriotic anthem and holiday. It recalls old martial memories. A nation's patriotic history is solely the history of its wars, that is, of the State in its health and glorious functioning. So in responding to the appeal of the flag, we are responding to the appeal of the State, to the symbol of the herd organized as an offensive and defensive body, conscious of its prowess and its mystical herd strength.

Even those authorities in the present Administration, to whom has been granted autocratic control over opinion, feel, though they are scarcely able to philosophize over, this distinction. It has been authoritatively declared that the horrid penalties against seditious opinion must not be construed as inhibiting legitimate, that is, partisan criticism of the Administration. A distinction is made between the Administration and the Government. It is quite accurately suggested by this attitude that the Administration is a temporary band of partisan politicians in charge of the machinery of Government, carrying out the mystical policies of State. The manner in which they operate this machinery may be freely discussed and objected to by their political opponents. The Governmental machinery may also be legitimately altered, in case of necessity. What may not be discussed or criticized is the mystical policy itself or the motives of the State in inaugurating such a policy. The President, it is true, has made certain partisan distinctions between candidates for office on the ground of support or nonsupport of the Administration, but what he means was really support or nonsupport of the State policy as faithfully carried out by the Administration. Certain of the Administration measures were devised directly to increase the health of the State, such as the Conscription and the Espionage laws. Others were concerned merely with the machinery. To oppose the first was to oppose the State and was therefore not tolerable. To oppose the second was to oppose fallible human judgment, and was therefore, though to be depreciated, not to be wholly interpreted as political suicide.

The distinction between Government and State, however, has not been so carefully observed. In time of war it is natural that Government as the seat of authority should be confused with the State or the mystic source of authority. You cannot very well injure a mystical idea which is the State, but you can very well interfere with the processes of Government. So that the two become identified in the public mind, and any contempt for or opposition to the workings of the machinery of Government is considered equivalent to contempt for the sacred State. The State, it is felt, is being injured in its faithful surrogate, and public emotion rallies passionately to defend it. It even makes any criticism of the form of Government a crime.

The inextricable union of militarism and the State is beautifully shown by those laws which emphasize interference with the Army and Navy as the most culpable of seditious crimes. Pragmatically, a case of capitalistic sabotage, or a strike in war industry would seem to be far more dangerous to the successful prosecution of the war than the isolated and ineffectual efforts of an individual to prevent recruiting. But in the tradition of the State ideal, such industrial interference with national policy is not identified as a crime against the State. It may be grumbled against; it may be seen quite rationally as an impediment of the utmost gravity. But it is not felt in those obscure seats of the herd mind which dictate the identity of crime and fix their proportional punishments. Army and Navy, however, are the very arms of the State; in them flows its most precious lifeblood. To paralyze them is to touch the very State itself. And the majesty of the State is so sacred that even to attempt such a paralysis is a crime equal to a successful strike. The will is deemed sufficient. Even though the individual in his effort to impede recruiting should utterly and lamentably fail, he shall be in no wise spared. Let the wrath of the State descend upon him for his impiety! Even if he does not try any overt action, but merely utters sentiments that may incidentally in the most indirect way cause someone to refrain from enlisting, he is guilty. The guardians of the State do not ask whether any pragmatic effect flowed out of this evil will or desire. It is enough that the will is present. Fifteen or twenty years in prison is not deemed too much for such sacrilege.

Such attitudes and such laws, which affront every principle of human reason, are no accident, nor are they the result of hysteria caused by the war. They are considered just, proper, beautiful by all the classes which have the State ideal, and they express only an extreme of health and vigor in the reaction of the State to its nonfriends.

Such attitudes are inevitable as arising from the devotees of the State. For the State is a personal as well as a mystical symbol, and it can only be understood by tracing its historical origin. The modern State is not the rational and intelligent product of modern men desiring to live harmoniously together with security of life, property, and opinion. It is not an organization which has been devised as pragmatic means to a desired social end. All the idealism with which we have been instructed to endow the State is the fruit of our retrospective imaginations. What it does for us in the way of security and benefit of life, it does incidentally as a by-product and development of its original functions, and not because at any time men or classes in the full possession of their insight and intelligence have desired that it be so. It is very important that we should occasionally lift the incorrigible veil of that ex post facto idealism by which we throw a glamour of rationalization over what is, and pretend in the ecstasies of social conceit that we have personally invented and set up for the glory of God and man the hoary institutions which we see around us. Things are what they are, and come down to us with all their thick encrustations of error and malevolence. Political philosophy can delight us with fantasy and convince us who need illusion to live that the actual is a fair and approximate copy—full of failings, of course, but approximately sound and sincere—of that ideal society which we can imagine ourselves as creating. From this it is a step to the tacit assumption that we have somehow had a hand in its creation and are responsible for its maintenance and sanctity.

Nothing is more obvious, however, than that every one of us comes into society as into something in whose creation we had not the slightest hand. We have not even the advantage, like those little unborn souls in The Blue Bird, of consciousness before we take up our careers on earth. By the time we find ourselves here we are caught in a network of customs and attitudes, the major directions of our desires and interests have been stamped on our minds, and by the time we have emerged from tutelage and reached the years of discretion when we might conceivably throw our influence to the reshaping of social institutions, most of us have been so molded into the society and class we live in that we are scarcely aware of any distinction between ourselves as judging, desiring individuals and our social environment. We have been kneaded so successfully that we approve of what our society approves, desire what our society desires, and add to the group our own passionate inertia against change, against the effort of reason, and the adventure of beauty.

Every one of us, without exception, is born into a society that is given, just as the fauna and flora of our environment are given. Society and its institutions are, to the individual who enters it, as much naturalistic phenomena as is the weather itself. There is, therefore, no natural sanctity in the State any more than there is in the weather. We may bow down before it, just as our ancestors bowed before the sun and moon, but it is only because something in us unregenerate finds satisfaction in such an attitude, not because there is anything inherently reverential in the institution worshiped. Once the State has begun to function, and a large class finds its interest and its expression of power in maintaining the State, this ruling class may compel obedience from any uninterested minority. The State thus becomes an instrument by which the power of the whole herd is wielded for the benefit of a class. The rulers soon learn to capitalize the reverence which the State produces in the majority, and turn it into a general resistance toward a lessening of their privileges. The sanctity of the State becomes identified with the sanctity of the ruling class, and the latter are permitted to remain in power under the impression that in obeying and serving them, we are obeying and serving society, the nation, the great collectivity of all of us. . . .

Appendix E

The War and the Intellectuals

By Randolph Bourne

From: *The Seven Arts*, 2 (June, 1917), 133-146

To those of us who still retain an irreconcilable animus against war, it has been a bitter experience to see the unanimity with which the American intellectuals have thrown their support to the use of war technique in the crisis in which America found herself. Socialists, college professors, publicists, new-republicans, practitioners of literature, have vied with each other in confirming with their intellectual faith the collapse of neutrality and the riveting of the war mind on a hundred million more of the world's people. And the intellectuals are not content with confirming our belligerent gesture. They are now complacently asserting that it was they who effectively willed it, against the hesitation and dim perceptions of the American democratic masses. A war made deliberately by the intellectuals! A calm moral verdict, arrived at after a penetrating study of inexorable facts! Sluggish masses, too remote from the world conflict to be stirred, too lacking in intellect to perceive their danger! An alert intellectual class saving the people in spite of themselves, biding their time with Fabian strategy until the nation could be moved into war without serious resistance! An intellectual class gently guiding a nation through sheer force of ideas into what the other nations entered only through predatory craft or popular hysteria or militarist madness! A war free from any taint of self-seeking, a war that will secure the triumph of democracy and internationalize the world! This is the picture which the more self-conscious intellectuals have formed of themselves, and which they are slowly impressing upon a population which is being led no man knows whither by an indubitably intellectualized President. And they are right, in that the war certainly did not spring from either the ideals or the prejudices, from the national ambitions or hysterias, of the American people, however acquiescent the masses prove to be, and however clearly the intellectuals prove their putative intuition.

Those intellectuals who have felt themselves totally out of sympathy with this drag toward war will seek some explanation for this joyful leadership. They will want to understand this willingness of the American intellect to open the sluices and flood us with the sewage of the war spirit. We cannot forget the virtuous horror and stupefaction which filled our college professors when they read the famous manifesto of their ninety-three German colleagues in defense of their war. To the American academic mind of 1914 defense of war was inconceivable. From Bernhardi it recoiled as from a blasphemy, little dreaming that two years later would find it creating its own cleanly reasons for imposing military service on the country and for talking of the rough rude currents of health and regeneration that war would send through the American body politic. They would have thought any one mad who talked of shipping American men by the hundreds of thousands-conscripts-to die on the fields of France. Such a spiritual change seems catastrophic when we shoot our minds back to those days when neutrality was a proud thing. But the intellectual progress has been so gradual that the country retains little sense of the irony. The war sentiment, begun so gradually but so perseveringly by the preparedness advocates who came from the ranks of big business, caught hold of one after another of the intellectual groups. With the aid of Roosevelt, the murmurs became a monotonous chant, and finally a chorus so mighty that to be out of it was at first to be disreputable and finally almost obscene. And slowly a strident rant was worked up against Germany which compared very creditably with the German fulminations against the greedy power of England. The nerve of the war feeling centered, of course, in the richer and older classes of the Atlantic seaboard, and was keenest where there were French or English business and particularly social connections. The sentiment then spread over the country as a class phenomenon, touching everywhere those upper-class elements in each section who identified themselves with this eastern ruling group. It must never be forgotten that in every community it was the least liberal and least democratic elements among whom the preparedness and later the war sentiment was found. The farmers were apathetic, the small businessmen and workingmen are still apathetic toward the war. The election was a vote of confidence of these latter classes in a President who would keep the faith of neutrality. The intellectuals, in other words, have identified themselves with the least democratic forces in American life. They have assumed the leadership for war of those very classes whom the American democracy has been immemorially fighting. Only in a world where irony was dead could an intellectual class enter war at the head of such illiberal cohorts in the avowed cause of world liberalism and world democracy. No one is left to point out the undemocratic nature of this war liberalism. In a time of faith, skepticism is the most intolerable of all insults.

Our intellectual class might have been occupied, during the last two years of war, in studying and clarifying the ideals and aspirations of the American democracy, in discovering a true Americanism which would not have been merely nebulous but might have federated the different ethnic groups and traditions. They might have spent the time in endeavoring to clear the public mind of the cant of war, to get rid of old mystical notions that clog our thinking. We might have used the time for a great wave of education, for setting our house in spiritual order. We could at least have set the problem before ourselves. If our intellectuals were going to lead the administration, they might conceivably have tried to find some way of securing peace by making neutrality effective. They might have turned their intellectual energy not to the problem of jockeying the nation into war, but to the problem of using our vast neutral power to attain democratic ends for the rest of the world and ourselves without the use of the malevolent technique of war. They might have failed. The point is that they scarcely tried. The time was spent not in clarification and education, but in a mulling over of nebulous ideals of democracy and liberalism and civilization which had never meant anything fruitful to those ruling classes who now so glibly used them, and in giving free rein to the elementary instinct of self-defense. The whole era has been spiritually wasted. The outstanding feature has been not its Americanism but its intense colonialism. The offense of our intellectuals was not so much that they were colonial-for what could we expect of a nation composed of so many national elements?—but that it was so one-sidedly and partisanly colonial. The official, reputable expression of the intellectual class has been that of the English colonial. Certain portions of it have been even more loyalist than the king, more British even than Australia. Other colonial attitudes have been vulgar. The colonialism of the other American stocks was denied a hearing from the start. America might have been made a meeting ground for the different national attitudes. An intellectual class, cultural colonists of the different European nations, might have threshed out the issues here as they could not be threshed out in Europe. Instead of this, the English colonials in university and press took command at the start, and we became an intellectual Hungary where thought was subject to an effective process of Magyarization. The reputable opinion of the American intellectuals became more and more either what could be read pleasantly in London, or what was written in an earnest effort to put Englishmen straight on their war aims and war technique. This Magyarization of thought produced as a counterreaction a peculiarly offensive and inept German apologetic, and the two partisans divided the field between them. The great masses, the other ethnic groups, were inarticulate. American public opinion was almost as little prepared for war in 1917 as it was in 1914.

The sterile results of such an intellectual policy are inevitable. During the war the American intellectual class has produced almost nothing in the way of original and illuminating interpretation. Veblen's Imperial Germany; Patten's Culture and War, and addresses; Dewey's German Philosophy and Politics; a chapter or two in Weyl's American World Policies—is there much else of creative value in the intellectual repercussion of the war? It is true that the shock of war put the American intellectual to an unusual strain. He had to sit idle and think as spectator, not as actor. There was no government to which he could docilely and loyally tender his mind as did the Oxford professors to justify England in her own eyes. The American's training was such as to make the fact of war almost incredible. Both in his reading of history and in his lack of economic perspective he was badly prepared for it. He had to explain to himself something which was too colossal for the modern mind, which outran any language or terms which we had to interpret it in. He had to expand his sympathies to the breaking point, while pulling the past and present into some sort of interpretative order. The intellectuals in the fighting countries had only to rationalize and justify what their country was already doing. Their task was easy. A neutral, however, had really to search out the truth. Perhaps perspective was too much to ask of any mind. Certainly the older colonials among our college professors let their prejudices at once dictate their thought. They have been comfortable ever since. The war has taught them nothing and will teach them nothing. And they have had the satisfaction, under the rigor of events, of seeing prejudice submerge the intellects of their younger colleagues. And they have lived to see almost their entire class, pacifists and democrats too, join them as apologists for the "gigantic irrelevance" of war.

We have had to watch, therefore, in this country the same process which so shocked us abroad—the coalescence of the intellectual classes in support of the military program. In this country, indeed, the socialist intellectuals did not even have the grace of their German brothers and wait for the declaration of war before they broke for cover. And when they declared for war they showed how thin was the intellectual veneer of their socialism. For they called us in terms that might have emanated from any bourgeois journal to defend democracy and civilization, just as if it was not exactly against those very bourgeois democracies and capitalist civilizations that socialists had been fighting for decades. But so subtle is the spiritual chemistry of the "inside" that all this intellectual cohesion—herd instinct become her[d] intellect—which seemed abroad so hysterical and so servile, comes to us here in highly rational terms. We go to war to save the world from subjugation! But the German intellectuals went to war to save their culture from barbarization! And the French went to war to save their beautiful France! And the English to save international honor! And Russia, most altruistic and self-sacrificing of all, to save a small state from destruction! Whence is our miraculous intuition of our moral spotlessness? Whence our confidence that history will not unravel huge economic and imperialist forces upon which our rationalizations float like bubbles? The Jew often marvels that his race alone should have been chosen as the true people of the cosmic God. Are not our intellectuals equally fatuous when they tell us that our war of all wars is stainless and thrillingly achieving for good?

An intellectual class that was wholly rational would have called insistently for peace and not for war. For months the crying need has been for a negotiated peace, in order to [a]void the ruin of a deadlock. Would not the same amount of resolute statesmanship thrown into intervention have secured a peace that would have been a subjugation of neither side? Was the terrific bargaining power of a great neutral ever used? Our war followed, as all wars follow, a monstrous failure of diplomacy. Shamefacedness should now be our intellectuals' attitude, because the American play for peace was made so little more than a polite play. The intellectuals have still to explain why, willing as they now are to use force to continue the war to absolute exhaustion, they were not willing to use force to coerce the world to a speedy peace.

Their forward vision is no more convincing than their past rationality. We go to war now to internationalize the world! But surely their League to Enforce Peace is only a palpable apocalyptic myth, like the syndicalists' myth of the "general strike." It is not a rational program so much as a glowing symbol for the purpose of focusing belief, of setting enthusiasm on fire for international order. As far as it does this it has pragmatic value; but as far as it provides a certain radiant mirage of idealism for this war and for a world order founded on mutual fear, it is dangerous and obnoxious. Idealism should be kept for what is ideal. It is depressing to think that the prospect of a world so strong that none dare challenge it should be the immediate ideal of the American intel-

lectual. If the League is only a makeshift, a coalition into which we enter to restore order, then it is only a description of existing fact, and the
idea should be treated as such. But if it is an actually prospective outcome of the settlement, the keystone of American policy, it is neither
realizable nor desirable. For the program of such a League contains no
provision for dynamic national growth or for international economic
justice. In a world which requires recognition of economic internationalism far more than of political internationalism, an idea is reactionary
which proposes to petrify and federate the nations as political and economic units. Such a scheme for international order is a dubious justification for American policy. And if American policy had been sincere in
its belief that our participation would achieve international beatitude,
would we not have made our entrance into the war conditional upon a
solemn general agreement to respect in the final settlement these principles of international order? Could we have afforded, if our war was to
end war by the establishment of a league of honor, to risk the defeat of
our vision and our betrayal in the settlement? Yet we are in the war, and
no such solemn agreement was made, nor has it ever been suggested.

The case of the intellectuals seems, therefore, only very speciously
rational. They could have used their energy to force a just peace or at
least to devise other means than war for carrying through American policy. They could have used their intellectual energy to ensure that our
participation in the war meant the international order which they wish.
Intellect was not so used. It was used to lead an apathetic nation into an
irresponsible war, without guarantees from those belligerents whose
cause we were saving. The American intellectual, therefore, has been rational neither in his hindsight nor his foresight. To explain him we
must look beneath the intellectual reasons to the emotional disposition.
It is not so much what they thought as how they felt that explains our
intellectual class. Allowing for colonial sympathy, there was still the personal shock in a world war which outraged all our preconceived notions
of the way the world was tending. It reduced to rubbish most of the
humanitarian internationalism and democratic nationalism which had
been the emotional thread of our intellectuals' life. We had suddenly to
make a new orientation. There were mental conflicts. Our latent colonialism strove with our longing for American unity. Our desire for
peace strove with our desire for national responsibility in the world.
That first lofty and remote and not altogether unsound feeling of our
spiritual isolation from the conflict could not last. There was the itch to
be in the great experience which the rest of the world was having.
Numbers of intelligent people who had never been stirred by the horrors of capitalistic peace at home were shaken out of their slumber by

the horrors of war in Belgium. Never having felt responsibility for labor wars and oppressed masses and excluded races at home, they had a large fund of idle emotional capital to invest in the oppressed nationalities and ravaged villages of Europe. Hearts that had felt only ugly contempt for democratic strivings at home beat in tune with the struggle for freedom abroad. All this was natural, but it tended to overemphasize our responsibility. And it threw our thinking out of gear. The task of making our own country detailedly fit for peace was abandoned in favor of a feverish concern for the management of the war, advice to the fighting governments on all matters, military, social, and political, and a gradual working up of the conviction that we were ordained as a nation to lead all erring brothers toward the light of liberty and democracy. The failure of the American intellectual class to erect a creative attitude toward the war can be explained by these sterile mental conflicts which the shock to our ideals sent raging through us.

Mental conflicts end either in a new and higher synthesis or adjustment, or else in a reversion to more primitive ideas which have been outgrown but to which we drop when jolted out of our attained position. The war caused in America a recrudescence of nebulous ideals which a younger generation was fast outgrowing because it had passed the wistful stage and was discovering concrete ways of getting them incarnated in actual institutions. The shock of the war threw us back from this pragmatic work into an emotional bath of these old ideals. There was even a somewhat rarefied revival of our primitive Yankee boastfulness, the reversion of senility to that republican childhood when we expected the whole world to copy our republican institutions. We amusingly ignored the fact that it was just that Imperial German régime, to whom we are to teach the art of self-government, which our own federal structure, with its executive irresponsible in foreign policy and with its absence of parliamentary control, most resembles. And we are missing the exquisite irony of the unaffected homage paid by the American democratic intellectuals to the last and most detested of Britain's tory premiers as the representative of a "liberal" ally, as well as the irony of the selection of the best hated of America's bourbon "old guard" as the missionary of American democracy to Russia.

The intellectual state that could produce such things is one where reversion has taken place to more primitive ways of thinking. Simple syllogisms are substituted for analysis; things are known by their labels, our heart's desire dictates what we shall see. The American intellectual class, having failed to make the higher syntheses, regresses to ideas that can issue in quick, simplified action. Thought becomes any easy rationalization of what is actually going on or what is to happen inevitably tomorrow. It is true that certain groups did rationalize their colonialism and attach the doctrine of the inviolability of British sea power to the doctrine of a League of Peace. But this agile resolution of the mental conflict did not become a higher synthesis, to be creatively developed. It gradually merged into a justification for our going to war. It petrified into a dogma to be propagated. Criticism flagged and emotional propaganda began. Most of the socialists, the college professors and the practitioners of literature, however, have not even reached this high-water mark of synthesis. Their mental conflicts have been resolved much more simply. War in the interests of democracy! This was almost the sum of their philosophy. The primitive idea to which they regressed became almost insensibly translated into a craving for action. War was seen as the crowning relief of their indecision. At last action, irresponsibility, the end of anxious and torturing attempts to reconcile peace ideals with the drag of the world toward hell. An end to the pain of trying to adjust the facts to what they ought to be! Let us consecrate the facts as ideal! Let us join the greased slide toward war! The momentum increased. Hesitations, ironies, consciences, considerations—all were drowned in the elemental blare of doing something aggressive, colossal. The new-found Sabbath "peacefulness of being at war"! The thankfulness with which so many intellectuals lay down and floated with the current betrays the hesitation and suspense through which they had been. The American university is a brisk and happy place these days. Simple, unquestioning action has superseded the knots of thought. The thinker dances with reality.

With how many of the acceptors of war has it been mostly a dread of intellectual suspense? It is a mistake to suppose that intellectuality necessarily makes for suspended judgments. The intellect craves certitude. It takes effort to keep it supple and pliable. In a time of danger and disaster we jump desperately for some dogma to cling to. The time comes, if we try to hold out, when our nerves are sick with fatigue, and we seize in a great healing wave of release some doctrine that can be immediately translated into action. Neutrality meant suspense, and so it became the object of loathing to frayed nerves. The vital myth of the League of Peace provides a dogma to jump to. With war the world becomes motor again and speculation is brushed aside like cobwebs. The blessed emotion of self-defense intervenes too, which focused millions in Europe. A few keep up a critical pose after war is begun, but since they usually advise action which is in one-to-one correspondence with what the mass is already doing, their criticism is little more than a rationalization of the common emotional drive.

The results of war on the intellectual class are already apparent. Their thought becomes little more than a description and justification of what is going on. They turn upon any rash one who continues idly to speculate. Once the war is on, the conviction spreads that individual thought is helpless, that the only way one can count is as a cog in the great wheel. There is no good holding back. We are told to dry our unnoticed and ineffective tears and plunge into the great work. Not only is every one forced into line, but the new certitude becomes idealized. It is a noble realism which opposes itself to futile obstruction and the cowardly refusal to face facts. This realistic boast is so loud and sonorous that one wonders whether realism is always a stern and intelligent grappling with realities. May it not be sometimes a mere surrender to the actual, an abdication of the ideal through a sheer fatigue from intellectual suspense? The pacifist is roundly scolded for refusing to face the facts, and for retiring into his own world of sentimental desire. But is the realist, who refuses to challenge or criticize facts, entitled to any more credit than that which comes from following the line of least resistance? The realist thinks he at least can control events by linking himself to the forces that are moving. Perhaps he can. But if it is a question of controlling war, it is difficult to see how the child on the back of a mad elephant is to be any more effective in stopping the beast than is the child who tries to stop him from the ground. The ex-humanitarian, turned realist, sneers at the snobbish neutrality, colossal conceit, crooked thinking, dazed sensibilities, of those who are still unable to find any balm of consolation for this war. We manufacture consolations here in America while there are probably not a dozen men fighting in Europe who did not long ago give up every reason for their being there except that nobody knew how to get them away.

But the intellectuals whom the crisis has crystallized into an acceptance of war have put themselves into a terrifyingly strategic position. It is only on the craft, in the stream, they say, that one has any chance of controlling the current forces for liberal purposes. If we obstruct, we surrender all power for influence. If we responsibly approve, we then retain our power for guiding. We will be listened to as responsible thinkers, while those who obstructed the coming of war have committed intellectual suicide and shall be cast into outer darkness. Criticism by the ruling powers will only be accepted from those intellectuals who are in sympathy with the general tendency of the war. Well, it is true that they may guide, but if their stream leads to disaster and the frustration of national life, is their guiding any more than a preference whether they shall go over the right-hand or the left-hand side of the precipice? Meanwhile, however, there is comfort on board. Be with us,

they call, or be negligible, irrelevant. Dissenters are already excommuni-
cated. Irreconcilable radicals, wringing their hands among the debris,
become the most despicable and impotent of men. There seems no
choice for the intellectual but to join the mass of acceptance. But again
the terrible dilemma arises—either support what is going on, in which
case you count for nothing because you are swallowed in the mass and
great incalculable forces bear you on; or remain aloof, passively resis-
tant, in which case you count for nothing because you are outside the
machinery of reality.

Is there no place left, then, for the intellectual who cannot yet crys-
tallize, who does not dread suspense, and is not yet drugged with fatigue? The American intellectuals, in their preoccupation with reality,
seem to have forgotten that the real enemy is War rather than Imperial
Germany. There is work to be done to prevent this war of ours from
passing into popular mythology as a holy crusade. What shall we do
with leaders who tell us that we go to war in moral spotlessness, or who
make "democracy" synonymous with a republican form of government?
There is work to be done in still shouting that all the revolutionary by-
products will not justify the war, or make war anything else than the
most noxious complex of all the evils that afflict men. There must be
some to find no consolation whatever, and some to sneer at those who
buy the cheap emotion of sacrifice. There must be some irreconcilables
left who will not even accept the war with walrus tears. There must be
some to call unceasingly for peace, and some to insist that the terms of
settlement shall be not only liberal but democratic. There must be some
intellectuals who are not willing to use the old discredited counters
again and to support a peace which would leave all the old inflammable
materials of armament lying about the world. There must still be opposition to any contemplated "liberal" world order founded on military
coalitions. The "irreconcilable" need not be disloyal. He need not even
be "impossibilist." His apathy toward war should take the form of a
heightened energy and enthusiasm for the education, the art, the interpretation that make for life in the midst of the world of death. The intellectual who retains his animus against war will push out more boldly
than ever to make his case solid against it. The old ideals crumble; new
ideals must be forged. His mind will continue to roam widely and cease-
lessly. The thing he will fear most is premature crystallization. If the
American intellectual class rivets itself to a "liberal" philosophy that
perpetuates the old errors, there will then be need for "democrats"
whose task will be to divide, confuse, disturb, keep the intellectual wa-
ters constantly in motion to prevent any such ice from ever forming.

Appendix F

Adolf Hitler, to the assembly in the Reichstag on 01 October 1938, following the German invasion of Czechoslovakia.

For months the Germans in Sudetenland have been suffering under the torture of the Czechoslovak government. This is a problem which the Versailles Diktat created—a problem which has deteriorated until it becomes intolerable for us.

The Sudeten German population was and is a German. This German minority living there has been ill-treated in the most distressing manner. More than 1,000,000 people of German blood had in the years 1919-1920 to leave their homeland.

As always, I attempted to bring about, by the peaceful method of making proposals for revision, an alteration of this intolerable position. It is a lie when the outside world says that we only tried to carry through our revisions by pressure. Fifteen years before the National Socialist Party came to power there was the opportunity of carrying out these revisions by peaceful settlements and understanding. On my own initiative I have, not once but several times, made proposals for the revision of intolerable conditions. All these proposals, as you know, have been rejected—proposals for limitation of armaments and even, if necessary, disarmament, proposals for limitation of war making, proposals for the elimination of certain methods of modern warfare.

You know the proposals that I have made to fulfill the necessity of restoring German sovereignty over German territories. You know the endless attempts I made for a peaceful clarification and understanding of the problem of Austria and now the Sudetenland. It was all in vain.

Despite pressure from both us, Italy, France and Britain on Czechoslovakia to cede the Sudetenland to us, Beneš has as many times before refused to abide by the agreement, and the persecution of the Sudeten German minorities continue with his blessing.

This night the German population in Sudetenland was the victim of a massacre carried out by the Czechoslovak army which claimed the lives of 32 civilians.

Since 5:45am we have answered the call of the Sudeten Germans to assure their safety, and from now on terror will be met with terror.

Adolf Hitler speaking at the Reichstag on 01 October 1938. (Public domain.)

The oppressing regime in Prague must be stopped! Whoever fights with bombs will be fought with bombs. Whoever departs from the rules of humane warfare can only expect that we shall do the same.

I will continue this struggle, no matter against whom, until the safety of the Reich and its rights are secured. For five years now I have been working on the building up of the German defences. Over 90 million have in that time been spent on the building up of these defence forces. They are now the best equipped and are above all comparison with what they were in 1914. My trust in them is unshakable.

When I called up these forces and when I now ask sacrifices of the German people and if necessary every sacrifice, then I have a right to do so, for I also am today absolutely ready, just as we were formerly, to make every possible sacrifice. I am asking of no German man more than I myself was ready throughout four years at any time to do. There will be no hardships for Germans to which I myself will not submit. My whole life henceforth belongs more than ever to my people. I am from now on just first soldier of the German Reich. I have once more put on that coat that was the most sacred and dear to me. I will not take it off again until victory is secured, or I will not survive the outcome. Should anything happen to me in the struggle then my first successor is Party Comrade Göring; should anything happen to Party Comrade Göring my next successor is Party Comrade Hess. You would then be under obligation to give to them as Fuhrer the same blind loyalty and obedience as to myself. Should anything happen to Party Comrade Hess, then by law the Senate will be called, and will choose from its midst the most worthy—that is to say the bravest—successor.

As a National Socialist and as German soldier I enter upon this struggle with a stout heart. My whole life has been nothing but one long struggle for my people, for its restoration, and for Germany. There was only one watchword for that struggle: faith in this people. One word I have never learned: that is, surrender. If, however, anyone thinks that we are facing a hard time, I should ask him to remember that once a Prussian King, with a ridiculously small State, opposed a stronger coalition, and in three wars finally came out successful because that State had that stout heart that we need in these times. I would, therefore, like to assure the entire world that a November 1918 will never be repeated in German history. Just as I myself am ready at any time to stake my life—anyone can take it for my people and for Germany—so I ask the same of all others. Whoever, however, thinks he can oppose this national command, whether directly of indirectly, shall fall. We have nothing to do with traitors. We are all faithful to our old principle. It is quite unimportant whether we ourselves live, but it is essential that our people shall live, that Germany shall live.

The sacrifice that is demanded of us is not greater than the sacrifice that many generations have made. If we form a community closely bound together by vows, ready for anything, resolved never to surrender, then our will shall master every hardship and difficulty. And I would like to close with the declaration that I once made when I began the struggle for power in the Reich. I then said: "If our will is so strong that no hardship and suffering can subdue it, then our will and our German might shall prevail".

Appendix G

Remarks by President Barack Obama in His Address to the Nation on Libya.

National Defense University, Washington, D.C., 7:31— 7:58pm, EDT, 28 March 2011.

Tonight, I'd like to update the American people on the international effort that we have led in Libya—what we've done, what we plan to do, and why this matters to us.

I want to begin by paying tribute to our men and women in uniform who, once again, have acted with courage, professionalism and patriotism. They have moved with incredible speed and strength. Because of them and our dedicated diplomats, a coalition has been forged and countless lives have been saved.

Meanwhile, as we speak, our troops are supporting our ally Japan, leaving Iraq to its people, stopping the Taliban's momentum in Afghanistan, and going after al Qaeda all across the globe. As Commander-in-Chief, I'm grateful to our soldiers, sailors, airmen, Marines, Coast Guardsmen, and to their families. And I know all Americans share in that sentiment.

For generations, the United States of America has played a unique role as an anchor of global security and as an advocate for human freedom. Mindful of the risks and costs of military action, we are naturally reluctant to use force to solve the world's many challenges. But when our interests and values are at stake, we have a responsibility to act. That's what happened in Libya over the course of these last six weeks.

Libya sits directly between Tunisia and Egypt—two nations that inspired the world when their people rose up to take control of their own destiny. For more than four decades, the Libyan people have been ruled by a tyrant—Muammar Qaddafi. He has denied his people freedom, exploited their wealth, murdered opponents at home and abroad, and terrorized innocent people around the world—including Americans who were killed by Libyan agents.

U.S. President Barack Obama. (Source: Official photographic portrait in the public domain, under a Creative Commons License. Photo by Pete Souza, the official White House photographer.)

Last month, Qaddafi's grip of fear appeared to give way to the promise of freedom. In cities and towns across the country, Libyans took to the streets to claim their basic human rights. As one Libyan said, "For the first time we finally have hope that our nightmare of 40 years will soon be over".

Faced with this opposition, Qaddafi began attacking his people. As President, my immediate concern was the safety of our citizens, so we evacuated our embassy and all Americans who sought our assistance. Then we took a series of swift steps in a matter of days to answer Qaddafi's aggression. We froze more than $33 billion of Qaddafi's regime's assets. Joining with other nations at the United Nations Security Council, we broadened our sanctions, imposed an arms embargo, and enabled Qaddafi and those around him to be held accountable for their crimes. I made it clear that Qaddafi had lost the confidence of his people and the legitimacy to lead, and I said that he needed to step down from power.

In the face of the world's condemnation, Qaddafi chose to escalate his attacks, launching a military campaign against the Libyan people. Innocent people were targeted for killing. Hospitals and ambulances were attacked. Journalists were arrested, sexually assaulted, and killed. Supplies of food and fuel were choked off. Water for hundreds of thousands of people in Misurata was shut off. Cities and towns were shelled, mosques were destroyed, and apartment buildings reduced to rubble. Military jets and helicopter gunships were unleashed upon people who had no means to defend themselves against assaults from the air.

Confronted by this brutal repression and a looming humanitarian crisis, I ordered warships into the Mediterranean. European allies declared their willingness to commit resources to stop the killing. The Libyan opposition and the Arab League appealed to the world to save lives in Libya. And so at my direction, America led an effort with our allies at the United Nations Security Council to pass a historic resolution that authorized a no-fly zone to stop the regime's attacks from the air, and further authorized all necessary measures to protect the Libyan people.

Ten days ago, having tried to end the violence without using force, the international community offered Qaddafi a final chance to stop his campaign of killing, or face the consequences. Rather than stand down, his forces continued their advance, bearing down on the city of Benghazi, home to nearly 700,000 men, women and children who sought their freedom from fear.

At this point, the United States and the world faced a choice. Qaddafi declared he would show "no mercy" to his own people. He com-

pared them to rats, and threatened to go door to door to inflict punishment. In the past, we have seen him hang civilians in the streets, and
kill over a thousand people in a single day. Now we saw regime forces
on the outskirts of the city. We knew that if we wanted—if we waited
one more day, Benghazi, a city nearly the size of Charlotte, could suffer
a massacre that would have reverberated across the region and stained
the conscience of the world.

It was not in our national interest to let that happen. I refused to let
that happen. And so nine days ago, after consulting the bipartisan leadership of Congress, I authorized military action to stop the killing and
enforce U.N. Security Council Resolution 1973.

We struck regime forces approaching Benghazi to save that city and
the people within it. We hit Qaddafi's troops in neighboring Ajdabiya,
allowing the opposition to drive them out. We hit Qaddafi's air defenses, which paved the way for a no-fly zone. We targeted tanks and
military assets that had been choking off towns and cities, and we cut
off much of their source of supply. And tonight, I can report that we
have stopped Qaddafi's deadly advance.

In this effort, the United States has not acted alone. Instead, we
have been joined by a strong and growing coalition. This includes our
closest allies— nations like the United Kingdom, France, Canada, Den-
mark, Norway, Italy, Spain, Greece, and Turkey -- all of whom have
fought by our sides for decades. And it includes Arab partners like
Qatar and the United Arab Emirates, who have chosen to meet their
responsibilities to defend the Libyan people.

To summarize, then: In just one month, the United States has
worked with our international partners to mobilize a broad coalition,
secure an international mandate to protect civilians, stop an advancing
army, prevent a massacre, and establish a no-fly zone with our allies and
partners. To lend some perspective on how rapidly this military and
diplomatic response came together, when people were being brutalized
in Bosnia in the 1990s, it took the international community more than
a year to intervene with air power to protect civilians. It took us 31 days.

Moreover, we've accomplished these objectives consistent with the
pledge that I made to the American people at the outset of our military
operations. I said that America's role would be limited; that we would
not put ground troops into Libya; that we would focus our unique capabilities on the front end of the operation and that we would transfer
responsibility to our allies and partners. Tonight, we are fulfilling that
pledge.

Our most effective alliance, NATO, has taken command of the enforcement of the arms embargo and the no-fly zone. Last night, NATO

decided to take on the additional responsibility of protecting Libyan civilians. This transfer from the United States to NATO will take place on Wednesday. Going forward, the lead in enforcing the no-fly zone and protecting civilians on the ground will transition to our allies and partners, and I am fully confident that our coalition will keep the pressure on Qaddafi's remaining forces.

In that effort, the United States will play a supporting role ~ including intelligence, logistical support, search and rescue assistance, and capabilities to jam regime communications. Because of this transition to a broader, NATO-based coalition, the risk and cost of this operation —to our military and to American taxpayers—will be reduced significantly.

So for those who doubted our capacity to carry out this operation, I want to be clear: The United States of America has done what we said we would do.

That's not to say that our work is complete. In addition to our NATO responsibilities, we will work with the international community to provide assistance to the people of Libya, who need food for the hungry and medical care for the wounded. We will safeguard the more than $33 billion that was frozen from the Qaddafi regime so that it's available to rebuild Libya. After all, the money doesn't belong to Qaddafi or to us—it belongs to the Libyan people. And we'll make sure they receive it.

Tomorrow, Secretary Clinton will go to London, where she will meet with the Libyan opposition and consult with more than 30 nations. These discussions will focus on what kind of political effort is necessary to pressure Qaddafi, while also supporting a transition to the future that the Libyan people deserve ~ because while our military mission is narrowly focused on saving lives, we continue to pursue the broader goal of a Libya that belongs not to a dictator, but to its people.

Now, despite the success of our efforts over the past week, I know that some Americans continue to have questions about our efforts in Libya. Qaddafi has not yet stepped down from power, and until he does, Libya will remain dangerous. Moreover, even after Qaddafi does leave power, 40 years of tyranny has left Libya fractured and without strong civil institutions. The transition to a legitimate government that is responsive to the Libyan people will be a difficult task. And while the United States will do our part to help, it will be a task for the international community and—more importantly—a task for the Libyan people themselves.

In fact, much of the debate in Washington has put forward a false choice when it comes to Libya. On the one hand, some question why America should intervene at all— even in limited ways -- in this distant

land. They argue that there are many places in the world where innocent civilians face brutal violence at the hands of their government, and
America should not be expected to police the world, particularly when
we have so many pressing needs here at home.

It's true that America cannot use our military wherever repression
occurs. And given the costs and risks of intervention, we must always
measure our interests against the need for action. But that cannot be an
argument for never acting on behalf of what's right. In this particular
country—Libya—at this particular moment, we were faced with the prospect of violence on a horrific scale. We had a unique ability to stop that
violence: an international mandate for action, a broad coalition prepared to join us, the support of Arab countries, and a plea for help
from the Libyan people themselves. We also had the ability to stop
Qaddafi's forces in their tracks without putting American troops on the
ground.

To brush aside America's responsibility as a leader and—more pro-
foundly—our responsibilities to our fellow human beings under such
circumstances would have been a betrayal of who we are. Some nations
may be able to turn a blind eye to atrocities in other countries. The
United States of America is different. And as President, I refused to
wait for the images of slaughter and mass graves before taking action.

Moreover, America has an important strategic interest in preventing
Qaddafi from overrunning those who oppose him. A massacre would
have driven thousands of additional refugees across Libya's borders,
putting enormous strains on the peaceful -- yet fragile— transitions in
Egypt and Tunisia. The democratic impulses that are dawning across
the region would be eclipsed by the darkest form of dictatorship, as repressive leaders concluded that violence is the best strategy to cling to
power. The writ of the United Nations Security Council would have
been shown to be little more than empty words, crippling that institution's future credibility to uphold global peace and security. So while I
will never minimize the costs involved in military action, I am convinced that a failure to act in Libya would have carried a far greater
price for America.

Now, just as there are those who have argued against intervention in
Libya, there are others who have suggested that we broaden our military
mission beyond the task of protecting the Libyan people, and do whatever it takes to bring down Qaddafi and usher in a new government.

Of course, there is no question that Libya—and the world—would be
better off with Qaddafi out of power. I, along with many other world
leaders, have embraced that goal, and will actively pursue it through

non-military means. But broadening our military mission to include regime change would be a mistake.

The task that I assigned our forces—to protect the Libyan people from immediate danger, and to establish a no-fly zone—carries with it a U.N. mandate and international support. It's also what the Libyan opposition asked us to do. If we tried to overthrow Qaddafi by force, our coalition would splinter. We would likely have to put U.S. troops on the ground to accomplish that mission, or risk killing many civilians from the air. The dangers faced by our men and women in uniform would be far greater. So would the costs and our share of the responsibility for what comes next.

To be blunt, we went down that road in Iraq. Thanks to the extraordinary sacrifices of our troops and the determination of our diplomats, we are hopeful about Iraq's future. But regime change there took eight years, thousands of American and Iraqi lives, and nearly a trillion dollars. That is not something we can afford to repeat in Libya.

As the bulk of our military effort ratchets down, what we can do—and will do—is support the aspirations of the Libyan people. We have intervened to stop a massacre, and we will work with our allies and partners to maintain the safety of civilians. We will deny the regime arms, cut off its supplies of cash, assist the opposition, and work with other nations to hasten the day when Qaddafi leaves power. It may not happen overnight, as a badly weakened Qaddafi tries desperately to hang on to power. But it should be clear to those around Qaddafi, and to every Libyan, that history is not on Qaddafi's side. With the time and space that we have provided for the Libyan people, they will be able to determine their own destiny, and that is how it should be.

Let me close by addressing what this action says about the use of America's military power, and America's broader leadership in the world, under my presidency.

As Commander-in-Chief, I have no greater responsibility than keeping this country safe. And no decision weighs on me more than when to deploy our men and women in uniform. I've made it clear that I will never hesitate to use our military swiftly, decisively, and unilaterally when necessary to defend our people, our homeland, our allies and our core interests. That's why we're going after al Qaeda wherever they seek a foothold. That is why we continue to fight in Afghanistan, even as we have ended our combat mission in Iraq and removed more than 100,000 troops from that country.

There will be times, though, when our safety is not directly threatened, but our interests and our values are. Sometimes, the course of history poses challenges that threaten our common humanity and our common security—responding to natural disasters, for example; or preventing genocide and keeping the peace; ensuring regional security, and maintaining the flow of commerce. These may not be America's problems alone, but they are important to us. They're problems worth solving. And in these circumstances, we know that the United States, as the world's most powerful nation, will often be called upon to help.

In such cases, we should not be afraid to act—but the burden of action should not be America's alone. As we have in Libya, our task is instead to mobilize the international community for collective action. Because contrary to the claims of some, American leadership is not simply a matter of going it alone and bearing all of the burden ourselves. Real leadership creates the conditions and coalitions for others to step up as well; to work with allies and partners so that they bear their share of the burden and pay their share of the costs; and to see that the principles of justice and human dignity are upheld by all.

That's the kind of leadership we've shown in Libya. Of course, even when we act as part of a coalition, the risks of any military action will be high. Those risks were realized when one of our planes malfunctioned over Libya. Yet when one of our airmen parachuted to the ground, in a country whose leader has so often demonized the United States—in a region that has such a difficult history with our country—this American did not find enemies. Instead, he was met by people who embraced him. One young Libyan who came to his aid said, "We are your friends. We are so grateful to those men who are protecting the skies".

This voice is just one of many in a region where a new generation is refusing to be denied their rights and opportunities any longer.

Yes, this change will make the world more complicated for a time. Progress will be uneven, and change will come differently to different countries. There are places, like Egypt, where this change will inspire us and raise our hopes. And then there will be places, like Iran, where change is fiercely suppressed. The dark forces of civil conflict and sectarian war will have to be averted, and difficult political and economic concerns will have to be addressed.

The United States will not be able to dictate the pace and scope of this change. Only the people of the region can do that. But we can make a difference.

I believe that this movement of change cannot be turned back, and that we must stand alongside those who believe in the same core principles that have guided us through many storms: our opposition to violence directed at one's own people; our support for a set of universal rights, including the freedom for people to express themselves and choose their leaders; our support for governments that are ultimately responsive to the aspirations of the people.

Born, as we are, out of a revolution by those who longed to be free, we welcome the fact that history is on the move in the Middle East and North Africa, and that young people are leading the way. Because wherever people long to be free, they will find a friend in the United States. Ultimately, it is that faith—those ideals—that are the true measure of American leadership.

My fellow Americans, I know that at a time of upheaval overseas—when the news is filled with conflict and change—it can be tempting to turn away from the world. And as I've said before, our strength abroad is anchored in our strength here at home. That must always be our North Star—the ability of our people to reach their potential, to make wise choices with our resources, to enlarge the prosperity that serves as a wellspring for our power, and to live the values that we hold so dear.

But let us also remember that for generations, we have done the hard work of protecting our own people, as well as millions around the globe. We have done so because we know that our own future is safer, our own future is brighter, if more of mankind can live with the bright light of freedom and dignity.

Tonight, let us give thanks for the Americans who are serving through these trying times, and the coalition that is carrying our effort forward. And let us look to the future with confidence and hope not only for our own country, but for all those yearning for freedom around the world.

Thank you. God bless you, and may God bless the United States of America. (Applause.) Thank you.

Appendix H

Congressman Dennis Kucinich's Address to Congress on the War in Libya, United States House of Representatives, 31 March 2011

Mr. Speaker.

The critical issue before this nation today is not Libyan democracy, it is American democracy. In the next hour I will describe the dangers facing our own democracy. The principles of world democracy are embodied in the UN Charter, conceived to end the scourge of war for all time. The hope that nations could turn their swords into plowshares reflects the timeless impulse of humanity for enduring peace and with it an enhanced opportunity to pursue happiness.

We are not naïve about the existence of forces in the world which work against peace and against human security, but it is our fervent wish that we shall never become like those whom we condemn as lawless and without scruples. For it is our duty as members of a democratic society to provide leadership by example, to not only articulate the highest standards but to walk down the path to peace and justice with those standards as our constant companions. Our moral leadership in the world depends chiefly upon the might and light of truth and not shock and awe, and ghastly glow of our 2,000 lb bombs.

Our dear nation stands at a crossroads. The direction we take will determine not what kind of nation we are but what kind of nation we will become. Will we become a nation which plots in secret to wage war? Will we become a nation that observes our Constitution only in matters of convenience? Will we become a nation which destroys the unity of the world community painstakingly pieced together from the ruins of World War II, a war which itself followed a war to end all wars?

Now, once again we stand poised at a precipice—forced to the edge by an Administration which has thrown caution to the winds and our Constitution to the ground.

Dennis Kucinich, Member of the U.S. House of Representatives from Ohio's 10th district. (Source: Congressional portrait, in the public domain.)

It is abundantly clear from a careful reading of our Declaration of Independence that our nation was born from nothing less than the rebellion of the human spirit against the arrogance of power.

More than 200 years ago it was the awareness of the unchecked arrogance of King George III that led our Founders to deliberately and carefully balance our constitution by articulating the rights of Congress in Article I, as the primary check by our citizens against the dangers they foresaw for our republic. Our constitution was derived from the human and political experience of our Founders who were aware of what happens when one person takes it upon himself or herself to assume rights and privileges which place them above their fellow citizens.

"But where," asked Tom Paine in his famous tract, Common Sense, "...is the king of America? I'll tell you Friend, he reigns above, and doth not make havoc of mankind like the Royal of Britain....so far as we approve of monarchy, that in America the law is king. For as in absolute governments the king is law, so in free countries the law ought to be king; and there ought to be no other".

The power to declare war is firmly and explicitly vested in the Congress of the United States under Article I, Section 8 of the Constitution.

Let us make no mistake about it, dropping 2000 lb bombs and unleashing the massive firepower of our air force on the capital of a sovereign state is in fact an act of war and no amount of legal acrobatics can make it otherwise.

It is that same arrogance of power which the former Senator from Arkansas, J. William Fulbright, saw shrouded in the deceit which carried us into the abyss of the war in Vietnam. We determined we would never again see another Vietnam. It was the awareness of the unchecked power and arrogance of the executive which led Congress to pass the War Powers Act.

The Congress through the War Powers Act provided the executive with an exception to unilaterally respond only when the nation was in actual or imminent danger; to "repel sudden attacks".

Today we are in a constitutional crisis because our chief executive has assumed for himself powers to wage war which are neither expressly defined nor implicit in the Constitution, nor permitted under the War Powers Act.

This is a challenge not just to the Administration, but to Congress itself: The President has no right to wrest that fundamental power from Congress—and we have no right to cede it to him.

We, Members of Congress can no more absolve our president of his responsibility to obey this profound constitutional mandate then we can absolve ourselves of our failure to rise to the instant challenge that is before us today.

We violate our sacred trust to the citizens of the United States and our oath to uphold the constitution if we surrender this great responsibility and through our own inaction acquiesce in another terrible war.

We must courageously defend the oath that we took to defend the Constitution of the United States of America or we forfeit our right to participate in representative government.

How can we pretend to hold other sovereigns to fundamental legal principles through wars in foreign lands if we do not hold our own presidents to fundamental legal principles at home?

We are staring not only into the maelstrom of war in Libya, but also the code of behavior we are establishing today sets a precedent for the potential of evermore violent maelstroms ahead in Syria, Iran, and the horrifying chaos of generalized war throughout the Middle East. Our continued occupation of Iraq and Afghanistan makes us more vulnerable, not less vulnerable, to being engulfed in this generalized war.

In two years we have moved from President Bush's doctrine of preventive war to President Obama's assertion of the right to go to war without even the pretext of a threat to our nation.

This Administration is now asserting the right to go to war because a nation may threaten force against those who have internally taken up arms against it. Our bombs began dropping even before the UN's International Commission of Inquiry could verify allegations of murder of non-combatant civilians by the Gaddafi regime.

The Administration deliberately avoided coming to Congress and furthermore rejects the principle that Congress has any role in this matter. Yesterday we learned that 'The Administration would forge ahead with military action even if Congress passed a resolution constraining the mission.'

This is a clear and arrogant violation of our Constitution. This is war. Even a war launched for humanitarian reasons is still a war. And—only Congress can declare war.

We saw in the President's address to the nation on March 28, 2011 how mismatched elements are being hastily stitched together into a new war doctrine:

1. Executive privilege to wage war
2. War based on verbal threats
3. Humanitarian war
4. Preemptive war
5. Unilateral war

6. War for regime change

7. War against a nation whose government this Administration determines to be illegitimate

8. War authorized through the UN Security Council

9. War authorized through NATO and the Arab League

10. War requested by a rebel group against its despised government.

But not a word about coming to the representatives of the people in the United States Congress to make this decision. At this moment sailors and marines aboard the USS Bataan are headed to a position off the coast of Libya. The sons and daughters of our constituents put their lives on the line for this country. We owe it to them to challenge a misguided and illegal doctrine which could put their lives in great danger. For we have an obligation to protect them as they pledge to defend our nation.

The Administration's new war doctrine will lead not to peace, but to more war. It will stretch even thinner our military. In 2007 the Center for American Progress released a report on the effects of the wars in Iraq and Afghanistan and the multiple deployments on our Armed Forces. The report cited a lack of military readiness. It cited high levels of Post Traumatic Stress Disorder and suicide.

The report was released just before President Bush's surge in Iraq. Just one year after the surge in Afghanistan and after eight years in Iraq, the President commits an all volunteer army to another war of choice. If the criteria for military intervention in another country is government-sponsored violence and instability, over commitment of our military will be virtually inevitable and our national security will be undermined.

It is clear that the Administration planned a war against Libya at least a month in advance. But why? The President cannot say that Libya is an imminent or actual threat. He cannot say that war against Libya is in our vital interest. He cannot say that Libya had the intention or capability of attacking the United States. He has not claimed Libya had weapons of mass destruction to be used against us.

We are told our nation's role is limited, yet, at the same time, it is being expanded.

We have been told the administration does not favor military regime change, but then they tell us the war cannot end until Gaddafi is no longer the leader. Further, two weeks earlier the President signed a secret order for the CIA to assist the rebels who are trying to oust Gaddafi.

We are told that the burdens of the war in Libya would be shared by a coalition, but the United States is providing the bulk of the money, the armaments and the organizational leadership.

We are told that the President has legal authority for this war under the UN Security Council Resolution 1973. But this resolution specifically does not authorize any ground elements. Furthermore, the administration exceeded the mandate of the resolution by providing the rebels with air cover. Thus this war against Libya violated our Constitution and has even violated the very authority which the administration claimed was sufficient to take our country to war.

We are told the Gaddafi regime has been illegitimate for four decades. But we were not told that in 2003 the U.S. dropped sanctions against Libya. We were not told that Gaddafi, in an effort to ingratiate himself with the West in general and with America in particular, accepted a market-based economic program led by the very harsh structural adjustment remedies of the IMF and the World Bank. This led to the wholesale privatization of his state enterprises, contributing to unemployment in Libya rising above 20%. CNN reported on December 19, 2003 that Libya acknowledged having a nuclear program, pledged to destroy weapons of mass destruction and pledged to allow international inspections.

This was a decision which President George W. Bush has praised saying Gaddafi's actions "made our country and our world safer".

We are told that Gaddafi is in breach of UN Security Council Resolutions but now our own Secretary of State is reportedly considering arming the rebels, an act which would be a breach of the UN Security Council resolution which established an arms embargo.

We are told we went to war at the request of and with the support of the Arab League but the Secretary General of the Arab League, Amr Moussa began asking questions immediately after the imposition of the "No Fly Zone" stating that what was happening in Libya "differs from the aim of imposing a No Fly Zone....what we want is the protection of civilians and not the shelling of civilians".

Even the Secretary General of NATO, an organization which the United States founded and generally controls, expressed concern saying "We are not in Libya to arm people but to protect people".

Is this is truly a humanitarian intervention? What is humanitarian about providing to one side of a conflict the ability to wage war against the other side of a conflict, which will inevitably trigger a civil war turning Libya into a graveyard?

The Administration has told us they do not really know who the rebels are, but they are considering arming them nonetheless. The fact that they are even thinking about arming these rebels makes one think they know exactly who the rebels are.

While a variety of individuals and institutions may comprise the so called opposition in Libya, in fact one of the most significant organizations is the National Front for the Salvation of Libya (NFSL) along with its military front, the Libyan National Army. The NFSL's call for opposition to the Gaddafi regime in February was a catalyst of the conflict which precipitated the humanitarian crisis which is now used to justify our armed intervention.

But how spontaneous was this rebellion?

The Congressional Research Service in a 1987 analysis of the Libyan opposition wrote:

> "Over twenty opposition groups exist outside Libya. The most important in 1987 was the Libyan National Salvation Front (LNSF) formed in October 1981....The LNSF claimed responsibility for the daring attack on Gaddafi's headquarters at Bab al Aziziyah on May 8, 1984. Although the coup attempt failed and Gaddafi escaped unscathed, dissident groups claimed that some eighty Libyans, Cubans and East Germans perished".

Significantly the CRS cited various "sources" as early as 1984 which claimed ". . . the United States Central Intelligence Agency trained and supported the LNSF [Libyan National Salvation Front] before and after the May 8 operation".

By October 31, 1996, according to the BBC translation of Al-Hayat, an Arabic journal in London, a Colonel Khalifah Hiftar, who was the leader of the Libyan National Liberation Army, the armed wing of the LNSF was quoted as saying "force is the only effective method" in dealing with Gaddafi.

Move forward to March 26, 2011. The McClatchy Newspapers reported that the "new leader of Libya's opposition military, left for Libya two weeks ago", apparently around the same time that the President signed the covert operations order. The new leader spent the past two decades of his life in suburban Virginia where he had no visible means of support. His name: Colonel Khalifah Hiftar. One wonders when he planned his trip and who is his travel agency.

Congress needs to determine whether the United States, through previous covert support of the armed insurrection driven by the American-created NFSL, potentially helped create the humanitarian crisis was used to justify military intervention?

If we really want to understand how our constitutional prerogative for determining war and peace has been preempted by this Administration, it is important that Congress fully consider relevant events which may relate directly to the attack on Libya.

Consider this: On November 2, 2010 France and Great Britain signed a mutual defense treaty, which included joint participation in "Southern Mistral" a series of war games outlined in the bilateral agreement and surprisingly documented on a joint military web site established by France and Great Britain. Southern Mistral involved a long-range conventional air attack, called Southern Storm, against a dictatorship in a fictitious southern country called "Southland," in response to a pretend attack on France by "Southland". The joint military air strike was authorized by a pretend United Nations Security Council Resolution. The "Composite Air Operations" were planned for the period of March 21-25, 2011.

On March 20, 2011 the United States joined France and Great Britain in an air attack against Libya, pursuant to UN Security Council Resolution 1973.

Have the scheduled war games simply been postponed, or are they actually under way after months of planning, under the name of Operation Odyssey Dawn? Were opposition forces in Libya informed by the US, the UK or France about the existence of Southern Mistral/Southern Storm, which may have encouraged them to actions leading to greater repression and a humanitarian crisis? In short was this war against Gaddafi's Libya planned or a spontaneous response to the great suffering which Gaddafi was visiting upon his opposition? Congress has not even considered this possibility.

NATO, which has now taken over enforcement of the no-fly zone, has morphed from an organization which pledged mutual support to defend North Atlantic states from aggression in military operations reaching from Libya to the Chinese border in Afghanistan. We need to now ask what role the French Air Force General Abrial and current Supreme Allied Commander of NATO for Transformation may have played in the development of Operation Southern Storm and in discussions with the U.S. in the expansion of the UN Mandate into a NATO operation. What has been the role of the US African Command and Central Command in discussions leading up to this conflict? What did we know and when did we know it?

The United Nations Security Council process is at risk when its members are not fully informed of all the facts when they authorize a military operation. It is at risk from NATO which is usurping its mandate without specific authorization of Security Council Resolution 1973. The United States pays 25% of the military expense of NATO and NATO may be participating in the expansion of the UN mandate.

The United Nations relies not only on its moral authority, but on the moral cooperation of its member nations. If America exceeds its legal authority and determines to redefine international law, we journey away from an international moral order and into the amorality of power politics where the rule of force trumps the rule of law.

What are the fundamental principles at stake in America today?

First and foremost is our system of checks and balances built into the Constitution to ensure that important decisions of state are developed through mutual respect and shared responsibility in order to ensure that collective knowledge, indeed the collective wisdom of the people, is brought to bear. Two former Secretaries of State, James Baker and Warren Christopher have spoken jointly to the "importance of meaningful consultation between the President and Congress before the nation is committed to war".

Our nation has an inherent right to defend itself and a solemn obligation to defend the Constitution. From the Gulf of Tonkin in Vietnam to the allegations of Weapons of Mass Destruction in Iraq we have learned from bitter experience that the determination to go to war must be based on verifiable facts carefully considered.

Finally, civilian deaths are always to be regretted. But, we must understand from our own Civil War more than 150 years ago that nations must resolve their own conflicts and shape their own destiny internally.

However horrible those internal conflicts may be, these local conflicts can become even more dreadful if armed intervention in a civil war results in the internationalization of that conflict.

The belief that war is inevitable makes of war a self-fulfilling prophecy.

The United States, in this new and complex world wracked with great movements of masses to transform their own government, must itself be open to transformation, away from intervention, away from trying to determine the leadership of other nations, away from covert operations to try to manipulate events, and towards a rendezvous with those great principles of self-determination which gave us birth.

In a world which is interconnected and interdependent. In a world which cries out for human unity, we must call upon the wisdom of our namesake and Founding Father, George Washington, to guide us in the days ahead.

"The constitution vests the power of declaring war in Congress; therefore no offensive expedition of importance can be undertaken until after they shall have deliberated upon the subject and authorized such a measure".

Washington also had a wish for the future America: "My wish is to see this plague of mankind, war, banished from the earth".

CONTRIBUTORS

Laura Beach is currently completing a BA Double Major in Anthropology and Human Geography at Concordia University in Montreal, Quebec, Canada. Her primary study interests include environmental resource management, environmental justice, alternatives to industrialized food production, international development, neoliberalism and militarization. Laura is co-founder of the environmental organization TAPthirst and an avid environmental and social justice activist. She is a proud recipient of the *Medaille du Lieutenant-gouverneur pour la jeunesse* in recognition of academic achievement and community engagement. Laura is originally from South Gower, Ontario.

Jessica Cobran is currently doing a joint specialization in Anthropology and Sociology at Concordia University in Montreal, Quebec, Canada. She is originally from Montreal, but she also spent some of her youth in Toronto, Ontario, Canada. Her interests vary, but she enjoys studying topics regarding law, human and civil rights, and ethnic relations. She also has a passion for writing and looks forward to publish more works in the near future.

Maximilian C. Forte is the director of the New Imperialism seminar in the Department of Sociology and Anthropology, at Concordia University in Montreal, Quebec, Canada. He is also the volume editor and publisher of the New Imperialism series for Alert Press. He is an associate professor in anthropology, specializing in political anthropology, Indigenous resurgence, and visual and media ethnographies.

Sabrina M. Guerrieri currently holds a BA Specialization in Anthropology at Concordia University in Montreal, Quebec, Canada. Her primary study interests have tended to focus on the political and cultural conflicts of the Middle Eastern region, particularly that of the

Islamic Republic of Iran. Sabrina plans to pursue research on the political culture of Iranian youths at the Graduate level, whereby she will examine how art is used as a medium of resistance.

MacLean Hawley is an anthropology student at McGill University in Montreal, Quebec, Canada. His studies involve work on the senses, stone tool analysis, and (typically N. American) archaeology. MacLean plans on pursuing Anthropology, with a focus on Biological Anthropology and Experimental Anthropology. He is from originally Minneapolis, Minnesota, and only recently moved to Montreal.

Natalie Jansezian is currently a BA Major in Sociology at Concordia University in Montreal, Quebec, Canada. Her primary study interests have tended to focus on the criminal justice system and juvenile delinquency. Natalie plans to pursue journalism at a Masters level. She is originally from Vancouver, British Columbia.

Corey Seaton is a BA Major in Anthropology at Concordia University in Montreal, Quebec, Canada. His primary study interests have tended to focus on ethnic communities, human rights, and media studies. Corey plans on doing pursuing his Master's once he has graduated. Corey is from Montreal.

INDEX

A

B

C

D

E